John Newman

Notes on Cylinder Bridge Piers and the Well System of Foundations

Especially Written to Assist those Engaged in the Construction of Bridges, Quays, Docks, River Walls, Weirs, etc.

John Newman

Notes on Cylinder Bridge Piers and the Well System of Foundations
Especially Written to Assist those Engaged in the Construction of Bridges, Quays, Docks, River Walls, Weirs, etc.

ISBN/EAN: 9783337232078

Printed in Europe, USA, Canada, Australia, Japan

Cover: Foto ©berggeist007 / pixelio.de

More available books at **www.hansebooks.com**

NOTES ON CYLINDER BRIDGE PIERS

AND THE

WELL SYSTEM OF FOUNDATIONS.

NOTES ON CYLINDER BRIDGE PIERS

AND THE

WELL SYSTEM OF FOUNDATIONS.

———◆———

Especially written to assist those engaged in the construction of

Bridges, Quays, Docks, River=Walls, Weirs, etc.

BY

JOHN NEWMAN,
Assoc.M.Inst.C.E., F.Impl.Inst.,

AUTHOR OF

"*Notes on Concrete and Works in Concrete*";
"*Earthwork Slips and Subsidences upon Public Works*";
"*Scamping Tricks and Odd Knowledge occasionally Practised upon Public Works*"; *etc., etc., etc.*

LONDON :
E. & F. N. SPON, 125, STRAND.

NEW YORK :
SPON & CHAMBERLAIN, 12, CORTLANDT STREET.

—

1893.

PREFACE.

THIS book has been especially written to assist those engaged in designing or erecting Cylinder Bridge Piers and Abutments, and Concrete, Brick, or Masonry Wells, as applied to Bridges, Quay-, Dock-, and River-Walls, etc. Many of the chapters have recently appeared as serial articles in THE ENGINEERING REVIEW.

It will be seen, by reference to the Table of Contents and the Index, that most of the chief points requiring attention in the design, sinking, or erection of Cylinder Piers or Wells, either by compressed air, dredging, or open-air excavation, from the first sketch and calculation to the completion of the work, are examined. The strains caused by wind-pressure on bridge piers, or the lateral thrust of earth on abutments and walls, are only very cursorily referred to, as there are many excellent treatises and papers on those subjects, whereas information on the matters herein mentioned is only to be fragmentarily obtained, and after considerable research in the various engineering journals, books, and reports of this and other countries, and especially in the engineering press.

In 1873 a Miller prize was awarded to the author by the Council of the Institution of Civil Engineers, for a short paper upon the calculations necessary in designing Iron Cylinder Bridge Piers, it being afterwards published by permission. The pamphlet having been many years out of print, and several engineers and bridge-builders, here and abroad, having unsolicitedly testified to their having received "much help" from it, the whole subject has been considered *de novo;* and although this is by no means an exhaustive treatise, it being a kind of miniature cyclopædia on "Cylinder Bridge Piers and the Well System of Foundations," and as the application of cylinder and well foundations has since been much extended, the hope is cherished that the book may be equally useful to the Engineer, Bridge-Builder, Contractor, and Student.

J. N.

LONDON, 1893.

CONTENTS.

PAGES

CHAPTER I.
GENERAL DESIGN... ... ... 1—10

CHAPTER II.
TO DETERMINE THE REQUIRED DIAMETER OF A CYLINDER BRIDGE PIER... ... ... 10—24

CHAPTER III.
LOAD ON THE BASE ... ... 24—33

CHAPTER IV.
SURFACE FRICTION ... ... 33—39

CHAPTER V.
SINKING CYLINDERS; GENERAL NOTES ... ... 40—49

CHAPTER VI.
SINKING CYLINDERS; STAGING; FLOATING OUT ... ... 50—53

CHAPTER VII.
REMOVING OBSTRUCTIONS IN SINKING, AND "RIGHTING" CYLINDERS ... ... 53—57

CHAPTER VIII.
KENTLEDGE ... ... 58—65

CHAPTER IX.
HEARTING ... ... 65—69

CHAPTER X.
The Compressed-Air Method of Sinking Cylinders ... 70—72

CHAPTER XI.
Limiting Depth; Air Supply and Leakage ... ... 73—80

CHAPTER XII.
Effects of Compressed Air on Men ... ... 80—83

CHAPTER XIII.
Air Locks... ... ... 83—87

CHAPTER XIV.
Working-Chamber, and Method of Lighting it ... ... 87—90

CHAPTER XV.
Excavating and Dredging Apparatus for Removing the Earth from the Interior of a Cylinder or Well 90—103

CHAPTER XVI.
Notes on some Dredging Apparatus used in Sinking Bridge Cylinders and Wells ... ... 103—111

CHAPTER XVII.
Sand-Pumps, Suction, Compressed-Air, and Water-Jet Dredgers ... ... 111—118

CHAPTER XVIII.
The Well System of Foundations for Bridge Piers, Abutments, Quays, and Dock Walls, etc. ... ... 118—129

CYLINDER BRIDGE PIERS.

CHAPTER I.

GENERAL DESIGN.

IN this book purely theoretical questions will not be specially examined, the object being to practically explain the chief points requiring consideration in the correct design of cylinder bridge piers and the well system of foundations, and in the prosecution of the sinking operations connected therewith. Reference will also be made to the load upon the base, surface friction, methods of sinking, and the general operations necessary in the design and erection of bridge piers or wells constructed according to the methods herein mentioned.

First, it may be stated as an axiom that no system of bridge piers or foundations can be universally recommended, because of the varying nature and condition of the ground and the different general circumstances. The cylinder pier system is usually employed where great lateral stability is not required; it is especially adapted for an insistent weight, and where a heavy load has to be supported without materially obstructing a river or waterway. It is obviously safer and cheaper to give too much waterway than too little; but economy of space in navigable rivers and rapid tideways is generally absolutely imperative in piers, both during erection and when erected; therefore, apart from other questions, the advantage of the cylinder method of foundations is apparent.

In deciding whether to use well foundations instead of iron cylinders filled with concrete, brickwork, or masonry, several questions must be taken into consideration, and among others may be named the following : The character of the soil, which should be sand or loose strata not firmer than sand; the probability of *débris*, and boulders, and other obstructions such as a hard stratum being encountered, in which case it may be advisable to adopt iron cylinders or the caisson system; the relative cost of the various types of bridge pier, as it may so happen that iron is cheap when bricks or Portland cement are dear; the length of the season during which operations can be carried on; and the assured ease and rapidity of erection. If the pier has to be made in a swift current, iron cylinders are to be preferred to wells, as there may be difficulty with the joints. In compact soil the difficulty of sinking a comparatively blunt-ended cylinder, such as brick or concrete, on a curb, gives a decided preference for an iron cylinder with a fine-cutting edge.

Provided the shoe, or curb, in non-metallic cylinders be made of the necessary strength, and the cutting edge of sufficient sharpness to be able to penetrate the soil without bending, and provided the ground should not be of an unequal degree of hardness, and the steining be well bolted and bonded, there is no reason why, with due care, partly non-metallic cylinders should not succeed in all ordinary loose soil. The thin cutting edge and complete union of the several parts are the chief advantages of metallic cylinders over non-metallic; but, on the other hand, the non-metallic possess greater weight, and therefore do not require so much loading during sinking operations. A consideration of the merits and demerits of each system naturally suggests a combination of the two methods of construction, by a union of the thin cutting edge of the metallic cylinder with the weight of the non-metallic, the cutting edge being at the outside diameter of the well, and the steining supported on a plate stiffened and strutted to the iron ring, and the whole of the steining bolted to the iron curb, and bonded throughout; but it is well to remember that with the ordinary curb and usual construction of the steining, the well system is very likely to be unsuccessful in any soils except loose sand and strata of that nature, unless special dredging plant is employed and more than ordinary care taken in sinking the wells. In the preceding remarks on the well system and non-metallic cylinders, they are not assumed to be sunk by the pneumatic method.

In situations where there is great packing of ice in the rivers, as in North Russia, Canada, and North America, bridges are not built on simple iron-cylinder piers, although these have been used in combination with the crib and other systems, because they do not afford sufficient lateral stability and weight to resist the packing of ice against them, and the severe blows which they would receive from large masses of floating ice; but they have been found most economical for river bridges in the tropics.

A system adopted by Mr. T. W. Kennard, at the Buffalo Bridge, on the Great Western Railway of Canada—where, owing to the immense force of the ice, very massive piers were required—was as follows:— Their lower portions were composed partly of broken stone in timber cribwork, and partly of masonry placed within wrought-iron cylinders, these latter being inside the timber cells; above the water line the pier consisted of ashlar masonry. By these means a cheap and effective pier was obtained, the great expense of set masonry was obviated, there was

GENERAL DESIGN.

sufficient mass to resist the force of the ice, and the piers were erected much quicker than if constructed of ashlar. The top ring of the cylinder was enlarged, so as to meet the next adjacent ring. The cylinders were 12 ft. in diameter. The width of the cribwork was 20 ft. The cylinders were placed 15 ft. apart, from centre to centre, and formed a single row. The almost constant depth of water was 40 ft. Where timber is cheap and stone at hand, this system is economical, quick, and effective for rivers subject to ice-floes, and where great stability and weight are essential.

The difficulty of binding cylinders together when sunk in a loose soil, so that they may act in accordance, is a reason against the use of this system for the whole of a pier of an arched bridge, unless it is combined with other methods; but in a firm stratum, with ordinary precautions, there is no reason why they should not be built upon with security against both scour and movement from thrust. Formerly cylinders from about 6 to 10 ft. in diameter were considered the handiest sizes for sinking; but now they are used with economy and success up to a diameter of about 21 ft. Cylinder foundations can be sunk to great depths and in deep water ; they are, in such situations, generally to be preferred to brick, masonry, or concrete piers, having no casing or requiring the erection of temporary cofferdams. The iron casing only wants ordinary staging, and protects and stiffens a cylinder bridge-pier, and prevents lateral movement in the hearting.

With cylinders of Portland cement, one of the chief precautions to be taken is to keep the interior dry by making thoroughly sound and reliable joints between the blocks. The effects of any leakage of water through the casing must be considered, in order that the hearting may be preserved in good condition, and protected from the action of air and water.

In the case of cylinders of considerable height above the ground, the diameter of the column should be sufficiently large to give lateral stability; but as the higher the column the larger the base, it is generally sufficient to calculate the area of the base ; however, the reducing ring should not be too abrupt, because the strain on the hearting in that portion of the cylinder above the reducing ring should not much exceed that below it. The load upon the hearting at the top of the cylinder, at the point of enlargement, and at the base, should be calculated. The enlarged base afforded by the use of a conical ring is of general utility where the strata have no great bearing power ; but when the base of the cylinder rests on rock there is no reason why the column should not be of one diameter throughout, as the rock is able to bear as much compressive strain as the hearting.

The best position for the commencement of the enlargement, or tapering, is just above the ground or the bed of the river, as there is then the least possible obstruction to the river, and increased area of the base and surface are obtained. When the lengths of the rings are from 6 to 9 ft., the diameter of the lower edge of the reducing-ring is ordinarily from about 1·4 to 1·5 times the upper diameter, and the slope from about ¼ to ⅓ to 1. The conical reducing-piece sometimes has a vertical bearing on the hearting, which may be obtained by having on the base of the reducing-ring an internal disc with an opening equal in diameter to that of the top rings of the cylinder, the disc being stayed and strengthened by vertical ribs. If there is no special reason to the contrary, the cylinders should be placed immediately under the main girders, and they should be braced together either at or about their tops ; but care must be taken that the bracing is sufficiently high in a navigable or tidal river, so that a barge or vessel cannot be sunk by being held under it. Cast-iron arches are sometimes turned between the cylinders, the level of the crown of the arch nearly corresponding with that of the top of the column. When the cylinders are connected at their summits by a girder, and the main beams of the bridge are firmly attached by an adjusting expansion arrangement to the tops, and the height of the cylinder above the ground is not more than four times the diameter, only light bracing, if any, in ordinary situations is usually requisite ; but when the main beams rest on rollers bracing is required. Should the height of the cylinder above the ground be from five to eight times the diameter, strong bracing is necessary, and it should increase as the difference between the height and the diameter of the column becomes greater. The tops of the cylinders should be connected by horizontal beams.

The iron cylinder system of bridge foundations is not economical if many cylinders have to be sunk close together ; the most efficient employment of that method is where one cylinder is sufficient for one main beam of a bridge, and only two to four cylinders are required for one pier. It is generally adopted for foundations in deep water, and of considerable depth in the ground. The well system is economical in sand or silt, and where the water is of moderate depth, and when the depth is too great for the employment of compressed air, provided special plant and excavating apparatus is used. In a tidal river the iron cylinder casing might be omitted at about low-water or flood-level, as the masonry or brickwork can then be built up in the open, but it should only rest upon the hearting and not upon the iron rings. Where rock crops out on the surface, or nearly so, cylindrical foundations are suitable, but the surface of the rock must be levelled by divers or by other means, so that the cylinder may have a level bed ; and the system is good if a hard substratum is soon reached after penetrating the upper strata.

In an opening bridge care must be taken that sufficient transverse stability is given to the piers, so as to resist the motion of opening and closing ; therefore, should the pier be composed of cylinders, they must be arranged and braced accordingly. On the Boston and Providence Railroad, U.S.A., cylinders 6 ft. in diameter were sunk 10 ft. into the mud, and twelve piles were driven in the interior of the column 40 ft. into the mud, or 30 ft. below the cylinders; the interior was then filled with cement concrete, and increased bearing was thereby gained.

The cylinder system has been used in the following manner in order to shorten the span of a bridge. The diameter of a cylinder has a set off of about 1 ft. 6 in., or 2 ft. on each side, upon which are firmly fixed two inclined struts which support the girders, in addition to the cylinder which is carried up to the underside of the superstructure. The base of these inclined struts should be out of the reach of blows from barges or shipping, etc., therefore this method appears better adapted for unnavigable rivers or land piers, than for rivers with any traffic.

With regard to the best form for a pier with an iron casing, experience shows that the circular is to be preferred, unless there are special reasons to the contrary, not only because a better casting can be thereby obtained, but also on account of sinking operations, as it has been found that if the columns are of an oblong, square, or flat elliptical section, and the soil is not homogeneous, they assume in sinking an oblique direction, and are difficult to get down in an exactly vertical position. The cylindrical is also the best form for resisting internal pressure and collapse. One of the objections to all forms excepting the circular or elliptical, is that they only have long straight side walls to resist the pressure of the earth, and the various strains during sinking. In all soils likely to swell such as some of the clays, the circular form is the best ; and it is the strongest form for the amount of metal used.

In deciding upon the relative position of the cylinders on plan, it should be remembered that in sand and moist soils much difficulty has been experienced in sinking cylinders when they have been placed very close together, as they have a tendency to draw one towards the other. About 3 ft. should be the minimum distance between the surfaces of the cylinders ; and for considerable depths practice shows that in sand they should not be nearer together than from one-fourth to one-fifth of their diameter, the minimum distance being as before stated.

If the question arises whether one or two large cylinders should be used, instead of many smaller columns, experience seems decidedly to point to the former being preferable, as they can be sunk with much greater certainty and at a less proportionate cost than the smaller cylinders, and are not so liable to get out of the vertical in sinking.

In order to lessen the frictional adhesion in sinking, the cutting edge is sometimes swelled out a little larger than the other cylinder rings, by making the top of the commencement of the V-shaped cutting edge with an external projection of about ½ in. beyond the outside diameter of its upper part. The cutting ring is usually thicker than the other lengths of the cylinder, and is brought to a taper to facilitate the sinking. It is generally about ¼ in. to ½ in. thicker than the ring above; there is no use, however, in having a thicker casing, if strength alone is required, than 2½ in., as the strength of the metal per square inch decreases very considerably beyond a certain thickness, if the rings are cast by the ordinary method. The thickness of the cutting ring and edge should be regulated by the nature of the soil through which the cylinder is to be sunk, and by the character of the obstructions likely to be encountered. If boulders are probable, and the cutting edge has to be thrust through them, it should be proportionally strong. It is usually made from one-third to one-half the height of the other rings, being but seldom above 4 ft. 6 in. in height.

If cylinders have to be sunk by the compressed air system at great depths, as from 80 to 100 feet below water, there will be considerable difficulty in keeping the column air tight, and a strain of some moment will be brought upon the casing from the pressure of the air requisite for the expulsion of water at such a depth. The thickness of the ring must be sufficient to sustain the weight necessary for sinking, and the strains brought upon it during that operation, in addition to the internal pressure arising from the use of the compressed air system. The rings are always made thicker than theory demands to provide for possible defects and the natural porosity of the metal. The thicknesses of metal generally used range from 1 in. to 1¾ in. A very small thickness would suffice to keep the hearting in position until it is set, but during sinking operations the cylinder has to resist various strains, and before it is filled it has frequently to withstand the pressure of the water. If the compressed air system is adopted the cylinder is also subject to internal strain. The flange joints should therefore be broad, and be planed if the pneumatic method of sinking is used at a considerable depth, or be carefully packed so as to distribute the strain, and give a uniform bearing and a fit over the entire surface.

Cast iron cylinders 6 ft. 6 in. in diameter, 1¼ in. and 1½ in. in thickness, have been cast in 9 ft. lengths in one piece. They have also been cast of greater diameter, such as 10 ft. to 15 ft., in 6 ft. lengths. In cylinders of ordinary diameter, it is advisable to lessen the length of each ring and make them in one piece, thus obviating the necessity of vertical joint flanges, the weight of which and the horizontal flanges amounts to a considerable percentage of the total weight of the cylinder Before deciding upon the lengths of the cylinder rings, it is advisable to inquire the sizes that manufacturers will undertake to cast soundly without extra cost; the saving in weight by lessening the number of joints, and the augmented lateral strength and air and water tightness

GENERAL DESIGN.   7

thereby gained, may also compensate for the increased cost of the rings.

In having the rings cast in one piece for large diameters, although the vertical joints are not required, yet owing to the increased number of horizontal joints, consequent upon the diminished lengths, it is obvious there is a point when the greater number of horizontal joints will require the same amount of metal as if vertical joints had been adopted. Each particular case must determine whether it is advisable or cheaper to have the rings with or without vertical joints; at the same time it should not be forgotten that very large homogeneous castings are more difficult to obtain than moderately sized pieces. The rings are usually cast in 9 ft. lengths. Care should be taken that any ornamental caps placed on the top of the upper ring of the cylinder are from 2 in. to 3 in. at the least below the bearing plates of the girder to prevent them being crushed. It is almost always impossible to sink several cylinders so that their tops are all level, as the subsidence under a load is hardly ever uniform, or the strata exactly horizontal. The capital, or top making up ring, should therefore not be cast until the test load has been removed from the columns.

Delay in attaching the top making-up ring is, however, frequently inconvenient, for when the castings have to be shipped, months may elapse before they are delivered. To obviate this, the next ring to the making-up piece is now sometimes made of a height of from 2 to 5 ft., and an adjustable top or making-up ring, from 2 to 3 ft. in height, is provided of larger diameter than the outside diameter of the cylinder, so that it can be bolted to the lower ring in any position required, the bolt holes in the lower ring being made on the site.

The object of the cutting ring being to cause easy and vertical penetration, it is clear that its form should be suited to the earth it has to penetrate. A chisel-pointed cutting edge is perhaps the best. With the view of preventing cracking of the cutting ring consequent upon its encountering an obstruction in sinking, such as boulders, tree stumps, seams of rock, and to resist the various strains caused by unequal loading of the cylinder or resistance of the ground, it has been made of wrought iron because of its less liability to fracture from blows and its more uniform strength, but care should be taken that it is well strutted and stayed so that no deformation can take place. Cast iron rings, however, are to be preferred for all the other rings, as these are more easily and quickly bolted together and are cheaper.

As an example of the unequal strain a cutting ring may have to sustain, let us assume a cylinder to be 11 ft. in diameter, and $1\frac{1}{2}$ in. in thickness. The area of the cutting edge would be $(132 + 1\frac{1}{2}) \times 3\cdot1416 \times 1\frac{1}{2} = 629$ square inches. Taking the weight of the iron only in the cylinder at, say, 50 tons, and the kentledge at the high figure of

350 tons, or a total weight of 400 tons, if the load and resistance were equable over the surface in contact, the pressure would be $\frac{400}{625} = 0.64$ ton per square inch, a very light load on good cast iron. But, as the strain may be unequal, and the cutting edge only rest, for example, for 4 ft. of its entire length upon a boulder, strains of various kinds may be caused; and if the whole weight be concentrated upon the boulder the cutting edge would be subject to a strain of

$$\frac{400 \text{ tons}}{48 \times 1\frac{1}{2}} = 5.5 \text{ tons per square inch.}$$

Although it is improbable the ground upon which the boulder rested would sustain such a weight, even if jammed between rock, still the strains on the cutting edge may be very irregular and severe, and while one part is not even in contact with the ground, another may be heavily strained, and this without considering the effect of the direction of the load, but merely taking it as vertical and direct-acting and the cutting edge as flat. Hence, although there may be no apparent fault in the metal, the cracking of the cutting ring in boulder ground or soil of unequal character is not unlikely under such circumstances. As boulders frequently occur in shoals in the bed of a river, it is well to determine, in deciding upon the site of a bridge, whether a slight change in its position may not considerably increase or decrease the cost.

It is advisable to remember in fixing upon the diameter of a cylinder or well, that by increasing the diameter the length of the perimeter of the cylinder or the cutting edge is reduced as compared with the area, and also the frictional resistance in the same proportion, thus lessening relatively to the area of the cylinder the liability of the cutting edge meeting with an obstruction. If the cylinders have to be sunk to considerable depths it is more by chance that they can be sunk in their exact position, and therefore the diameter should be sufficiently large to admit of unavoidable deviation from the true position, and not less than 1 ft. to 2 ft. should be allowed for possible divergence.

It can be claimed that where the ground is of varying hardness, and boulders or inclined thin strata of rock have to be penetrated, cylinders of small diameter are to be preferred, because sinking operations must be suspended until the obstruction is removed, and also in uneven soil a cylinder may partly rest upon a firm foundation on one side and on soft yielding ground on the other; whereas, in using two or more small cylinders instead of one large one, each can be sunk to different depths until a firm foundation is reached; and where a rock bed has a considerable inclination, or in perhaps the worst case that may occur, namely, when the cutting ring reaches a dipping stratum and the ground on the lower side of the cylinder is softer than the upper, then the cylinder is being pressed towards the soft lower side and may soon become slanting unless prompt measures are taken to counteract the pressure and produce equilibrium, and this is easier to accomplish in small cylinders. However, obstructions can be better removed in cylinders of large diameter, as methods of treatment can be adopted that cannot be used in a confined space. On the whole, the balance of advantage rests with large cylinders, but each system is likely to have advocates except in ground of a homogeneous nature, when undoubtedly large cylinders are to be preferred for the reasons previously named, and particularly for railway bridges, because they are more massive, and therefore better able to withstand not only a sudden and unequal rolling load, but also the horizontal thrust caused by the application of quick-acting continuous brakes in retarding or stopping a train upon a structure.

The joints of cylinders can be caulked with iron rust cement for half of the outer thickness of the rings ; and the space upon the inner half can be filled either with neat Portland cement, or one of fine sharp sand to one of Portland cement, so as to make it practically air-tight should it be expected that the pneumatic process of sinking may have to be used. In order to allow for concrete in the hearting swelling during the process of setting, or for unequal contraction or expansion of iron and the material in the cylinder, or freezing of water, the cylinders can be lined with tarred felt.

Cylinders with vertical sides are to be preferred to those with a splayed or trumpet-shaped end, as they are more likely to sink evenly and vertically, because they do not offer so much surface and resistance to any obstruction, nor do they disturb so much ground or impair any guidance that may be received from the earth, although, even if parallel sides are adopted, it by no means follows that cylinders will sink vertically.

In designing any temporary works, such as staging, care should be taken, especially in a soft river bed of mud or silt, that they are not unequally weighted, or the ground may be forced in one direction, causing undue pressure upon, or a run of soil into the cylinder.

The sinking operations connected with it should be duly considered in determining the general form of a cylinder and the manner of the weighting, which will be hereafter referred to under a separate head. Particularly in loose soil, experience shows that a fully weighted cylinder sinks quicker with fewer "blows" of soil into the interior and less trouble than one in which the loading is intermittent or comparatively light. With the view of utilising the permanent hearting of the cylinder for weighting the rings, an annular plate, made to support an internal ring of concrete, masonry or brickwork, is occasionally used, thus lessening the temporary load required for sinking operations, and keeping the centre of gravity of the cylinder lower than when its top is

temporarily loaded. In deciding upon the thickness of this annular ring of the hearting, sufficient working space must be left, so that excavating operations can be carried on with ease; perhaps the best material for such casing is Portland cement concrete, as it can be so made that it will fill the spaces between the flanges, ribs, feathers, lugs, and bolts, so as to leave no voids.

CHAPTER II.

To Determine the Required Diameter of a Cylinder Bridge Pier.

The following formulæ will give the required internal diameter of a cylinder bridge-pier, when the resistance from the frictional surface and the flotation power of the cylinder are disregarded :—

Let D = the required diameter of the cylinder in feet.

s = the safe load in tons per square foot upon the foundation.

W = the weight in tons of the superstructure on the cylinder, including the rolling load.

w = the weight in tons of the cylinder, including the hearting.

A = the required area of the foundation in square feet.

Then
$$A = \frac{W + w}{s},$$

and as the diameter of a circle = $1\cdot128 \sqrt{\text{area of circle}}$,

$$D = 1\cdot128 \sqrt{\left(\frac{W + w}{s}\right)}.$$

The value of w may be readily obtained by using the diagram Fig. 1, and that of W is known at the time of designing the pier.

Should any support from surface friction be taken into calculation, for ordinary depths in the ground, heights, and other conditions,

$$D \times 0\cdot75 \text{ to } D \times 0\cdot85$$

will approximately give the required diameter of the cylinder.

There are cogent practical reasons which prevent the frictional resistance being relied upon, and they will be hereafter named, unless it is certain that such surface friction cannot be disturbed or impaired.

The flotation power of the cylinder is not considered as a means of permanent support, because of its small value and mutability.

The diameters of the cylinder in the diagram of weights are given immediately below the horizontal base line, together with the mean

REQUIRED DIAMETER.

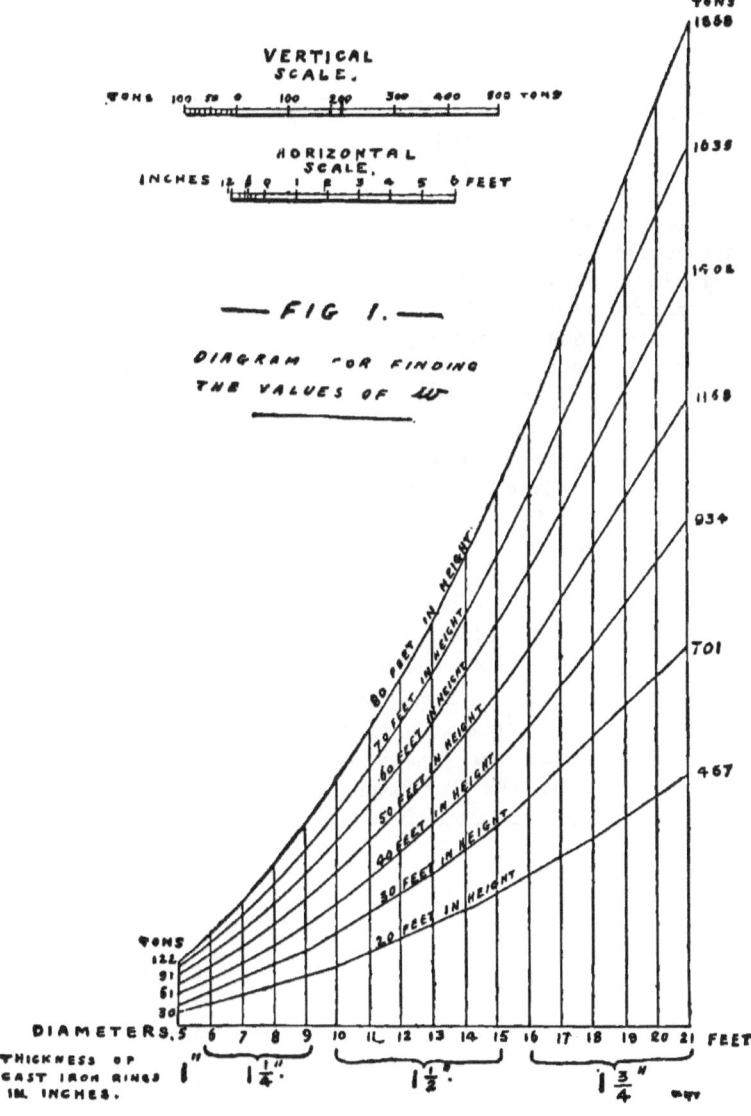

Fig. 1.—Diagram for finding the values of w.

Note.—*The weights in the diagram are calculated on the assumption that the hearting of the cylinder is Portland cement concrete. If brickwork, the weights will be about 16 per cent. less for picked stock bricks set in Portland cement mortar.*

thicknesses of the cast iron cylinder rings. An addition of 20 per cent. is made to the weight of the iron rings to allow for joint-flanges, bolts, lugs, bosses, etc.

Should the weight of a 10 ft. 6 in. cylinder be required, 45 ft. in height, the vertical line upon which the scale must be placed is midway between the 10 ft. and 11 ft. vertical lines. The required height is half-way between the 40 ft. and 50 ft. in height curved line, and in like manner any other dimensions or heights may be scaled. The diagram, which in other respects is self-explanatory, was made with the particular object of quickly and easily obtaining the value of w for the preliminary calculation of the required diameter of a cylinder bridge-pier, and for this purpose the weight of a wrought-iron cutting ring may be considered to be that of a cast-iron one.

Some formulæ are appended that may be found useful in calculating the required diameter of an iron cylinder bridge-pier :—

Let D = the internal diameter of the cylinder in feet.

W = the weight in tons of Portland cement concrete in the cylinder per lineal foot of the height of the cylinder, if the weight of Portland cement concrete is taken at 136 lbs. per cubic foot.

$$W = D^2 \times 0.048.$$

Let B = the weight of brickwork in Portland cement mortar, calculated at 112 lbs. per cubic foot, per lineal foot of the height of the cylinder.

$$B = D^2 \times 0.040.$$

Let I = the approximate weight of cast iron in tons per lineal foot of the height of the cylinder, including an allowance of 20 per cent. for joint flanges, ribs, bosses, lugs or strengthening brackets, bolts, etc.

Thickness of cast iron in the cylinder rings.

Inches.		
¾	... ...	$I = D \times .048$
1	... ...	$I = D \times .064$
1¼	... ...	$I = D \times .080$
1½	... ...	$I = D \times .096$
1¾	... ...	$I = D \times .112$
2	... ...	$I = D \times .127$
2¼	... ...	$I = D \times .143$
2½	... ...	$I = D \times .158$

The following is an empirical rule, deduced from many examples, for ascertaining the preliminary value of D in the case of cylinders of moderate total height, with the reducing ring at about the ground line; and for spans between 60 ft. and 200 ft., when two cylinders are used for a single line of railway, and a weight of 5 tons per square foot is taken as the safe load on the foundation, and the support from surface friction is disregarded :—

Let D = the internal diameter of the subterranean portion of a cylinder bridge-pier.

d = the internal diameter of the cylinder above the ground, or reducing ring.

s = span.

Then $$D = \sqrt{s}.$$

In the above rule $$d = \frac{D}{1 \cdot 4 \text{ to } 1 \cdot 5}.$$

If the safe load on the base is taken at, say, 6 tons per square foot instead of 5 tons, $\sqrt{(D^2 \times \frac{5}{6})}$ must be taken, and so in proportion for any other coefficient of the safe load on the foundation per square foot.

If support from surface friction is to be taken into calculation, then D, approximately $= 0 \cdot 8 \sqrt{s}$.

As in girders there is a limiting span, so in cylinder bridge-piers there is a limiting height for every diameter, beyond which the weight of the cylinder without any load will exceed the safe strain that the base or foundation will bear.

Let s = the safe load, in tons, on the foundation per square foot.

A = the area of the base in square feet.

W = the weight, in tons, on the cylinder from the superstructure, and the rolling load.

w = the weight of the cylinder per foot of height.

D = the diameter of the cylinder below the reducing ring, in feet.

d = the diameter, in feet, of the cylinder above the reducing ring.

H = the limiting height of the cylinder, in feet, measured from the base.

Then $$H = \frac{(s \times A) - W}{w}.$$

EXAMPLE—

Let s = 5 tons per square foot.

D = 12 feet ∴ A = 113·10 square feet.

d = 8 feet.

W = 80 tons.

Respecting the value of w, it would not be economical, or in accordance with the principles of correct design, for the cylinder to be of the same diameter throughout, excepting on rock foundations, when the safe load on the base approaches that which the hearting will safely bear. It is impossible to give any rule as to the exact position of the reducing ring; but in the case of ordinary foundations and conditions, by assuming that the portion of the cylinder sunk into the ground is one-third of the total height of the column, and that $D = 1\cdot5\,d$, w, in this example, will equal 4 in. in height of a 12 ft. cylinder, + 8 in. in height of an 8 ft. cylinder; $w \therefore = 5\cdot15$ tons, the hearting being Portland cement concrete.

$$H = \frac{(5 \times 113\cdot10)-80}{5\cdot15} = 94\cdot27 \text{ ft.}$$

If the diameter had been taken as the same throughout,

$$H = \frac{(5 \times 113\cdot10)-80}{8} = 60\cdot70 \text{ ft.}$$

A comparison of the two limiting heights will at once show the great advantage and economy of a reduction of the diameter of the cylinder above the ground-line.

The following is a calculation by aid of the diagram and the formulæ for the required diameter of the cylinders for a railway bridge :—

DATA.—Span 120 ft. Single line of railway of 4 ft. 8½ in. gauge. Two cylinders, each supporting one main beam. Cylinder to be sunk 30 ft. into the ground. Height of cylinder above the bed of river, 40 ft. Total height of cylinder, 70 ft. Reducing ring to commence at the bed of the river. Thickness of cylinder ring, 1½ in. Column to be filled with Portland cement concrete throughout. Safe load upon the foundation per square foot = 5 tons. Frictional resistance and flotation power not taken into account. Load on concrete per square foot not to exceed 7 tons. Rolling load to be calculated at 1¼ ton per lineal foot.

APPROXIMATE WEIGHT OF SUPERSTRUCTURE.

	Tons per lineal foot.
Girders, cross-girders, etc., 120 ft. span, weigh, say, 75 tons ... ... =	0·625
Roadway planking or floor ... ... =	0·161
Permanent way and ballast :—	
2 lineal ft. of rails at 72 lbs. per yard ... ... =	0·022
Fastenings ... ... =	0·002
Two longitudinal sleepers at 25 lbs. per foot ... =	0·022
Ballast. 16 ft. × 1 ft. × 3 in. thick = 4 cubic ft. × 150 lbs. = 600 lbs. ... ... } =	0·268
Total weight of superstructure per lineal foot ...	1·100

Rolling load ... ... 1·250

Total weight of superstructure and live load, in tons, per lineal foot ... ... 2·350

Weight of superstructure on one cylinder, rolling load included,
$$= \frac{120 \times 2\cdot35}{2} = 141 \text{ tons.}$$

Assume $D = \sqrt{\text{span}} = \sqrt{120} =$ say, 11 ft.

The cylinder is, therefore, 11 ft. in diameter for the 30 ft. in the ground.

Let $d =$ diameter of the cylinder above the ground,
$$d = \frac{11}{1\cdot5} = \text{say, 7 ft. 6 in.}$$

The cylinder is therefore 7 ft. 6 in. in diameter for the 40 ft. above the ground.

By diagram, the weights of the cylinder scale respectively:—

	Tons.
For the 11 ft. diameter, 30 ft. in height ... =	204
For the 7 ft. 6 in. diameter, 40 ft. in height ... =	132
	336

	Tons.
Superimposed load on one cylinder as before ... =	141
Weight of cylinder complete as above ... ... =	336
Total load on base	477

By formula:—
$$D = 1\cdot128 \sqrt{\left(\frac{W + w}{s}\right)},$$

$W = 141$ tons, $w = 336$ tons, $s = 5$ tons.

Therefore $D = 1\cdot128 \sqrt{\left(\frac{141 + 336}{5}\right)} = 11$ ft.

The portion of the cylinder in the ground is 11 ft. in diameter, therefore the area of the base $= 95\cdot04$ sq. ft., and the load on the base per square foot $= \frac{477}{95\cdot04} = 5\cdot02$ tons.

The normal load on the soil, which is assumed to be sand and to weigh ·055 ton a cubic foot, at a depth of 30 ft. $= 0\cdot055$ ton $\times 30 = 1\cdot65$ ton per square foot.

CYLINDER BRIDGE PIERS.

	Tons.
Total pressure on the base from the cylinder, superstructure, and rolling load, per square foot ... =	5·02
Normal pressure of the soil on the base, as above... =	1·65
Excess of pressure on the foundations in tons per square foot above the normal pressure ... =	3·37

The load on the concrete hearting per square foot at the commencement of the enlargement ring is as follows:—

The weight of concrete in a 7 ft. 6 in. cylinder is 8·05 tons per lineal yard.

	Tons.
Concrete 8·05 tons × 13¼ yards ... ... =	107·33
Superstructure and rolling load ... ... =	141·00
	248·33

The internal area of a 7 ft. 6 in. cylinder = 44·18 sq. ft.

The load on the concrete hearting per square foot $\frac{248\cdot33}{44\cdot18}$ = 5·62 tons.

The load from the concrete would be slightly less than this because the commencement of the reducing ring would probably be about 9 ft. above the ground, but this value is sufficiently near for all practical purposes. Should the 7 ft. 6 in. in diameter iron rings for this 40 ft. length be taken as bearing upon the concrete by means of the horizontal flanges, which should be the case, the strain on the concrete would be increased by 0·72 ton × 40 ft. = 28·80 tons, or an additional strain of

$$\frac{28\cdot80}{44\cdot18} = 0\cdot65 \text{ ton per square foot,}$$

making a total load of 5·62 + 0·65 = 6·27 tons per square foot, which strain is well within the safe limits of good Portland cement concrete properly mixed in the ordinary proportions.

If support from surface friction was relied upon, the required value of D would be about, D = 0·8 × 11 = say, 9 ft., and the diameter above the reducing ring, say, 7 ft.

	Tons.
Weight of a 9 ft. cylinder, 30 ft. in height ... =	137·40
Weight of a 7 ft. cylinder, 40 ft. in height ... =	115·92
	253·32
Weight of superstructure and rolling load, as before =	141·00
Total =	394·32

REQUIRED DIAMETER. 15

Rolling load ... ... 1·250

Total weight of superstructure and live load, in tons, per lineal foot ... ... 2·350

Weight of superstructure on one cylinder, rolling load included,

$$= \frac{120 \times 2\cdot 35}{2} = 141 \text{ tons.}$$

Assume $D = \sqrt{\text{span}} = \sqrt{120} =$ say, 11 ft.

The cylinder is, therefore, 11 ft. in diameter for the 30 ft. in the ground.

Let $d =$ diameter of the cylinder above the ground,

$$d = \frac{11}{1\cdot 5} = \text{say, 7 ft. 6 in.}$$

The cylinder is therefore 7 ft. 6 in. in diameter for the 40 ft. above the ground.

By diagram, the weights of the cylinder scale respectively:—

	Tons.
For the 11 ft. diameter, 30 ft. in height ... =	204
For the 7 ft. 6 in. diameter, 40 ft. in height ... =	132
	336

	Tons.
Superimposed load on one cylinder as before ... =	141
Weight of cylinder complete as above ... ... =	336
Total load on base	477

By formula:—

$$D = 1\cdot 128 \sqrt{\left(\frac{W + w}{s}\right)},$$

$W = 141$ tons, $w = 336$ tons, $s = 5$ tons.

Therefore $D = 1\cdot 128 \sqrt{\left(\frac{141 + 336}{5}\right)} = 11$ ft.

The portion of the cylinder in the ground is 11 ft. in diameter, therefore the area of the base $= 95\cdot 04$ sq. ft., and the load on the base per square foot $= \frac{477}{95\cdot 04} = 5\cdot 02$ tons.

The normal load on the soil, which is assumed to be sand and to weigh ·055 ton a cubic foot, at a depth of 30 ft. $= 0\cdot 055$ ton $\times 30 = 1\cdot 65$ ton per square foot.

CYLINDER BRIDGE PIERS.

	Tons.
Total pressure on the base from the cylinder, superstructure, and rolling load, per square foot ... =	5·02
Normal pressure of the soil on the base, as above... =	1·65
Excess of pressure on the foundations in tons per square foot above the normal pressure ... =	3·37

The load on the concrete hearting per square foot at the commencement of the enlargement ring is as follows:—

The weight of concrete in a 7 ft. 6 in. cylinder is 8·05 tons per lineal yard.

	Tons.
Concrete 8·05 tons × 13⅓ yards ... ... =	107·33
Superstructure and rolling load ... ... =	141·00
	248·33

The internal area of a 7 ft. 6 in. cylinder = 44·18 sq. ft.

The load on the concrete hearting per square foot $\frac{248\cdot33}{44\cdot18}$ = 5·62 tons.

The load from the concrete would be slightly less than this because the commencement of the reducing ring would probably be about 9 ft. above the ground, but this value is sufficiently near for all practical purposes. Should the 7 ft. 6 in. in diameter iron rings for this 40 ft. length be taken as bearing upon the concrete by means of the horizontal flanges, which should be the case, the strain on the concrete would be increased by 0·72 ton × 40 ft. = 28·80 tons, or an additional strain of

$$\frac{28\cdot80}{44\cdot18} = 0\cdot65 \text{ ton per square foot,}$$

making a total load of 5·62 + 0·65 = 6·27 tons per square foot, which strain is well within the safe limits of good Portland cement concrete properly mixed in the ordinary proportions.

If support from surface friction was relied upon, the required value of D would be about, D = 0·8 × 11 = say, 9 ft., and the diameter above the reducing ring, say, 7 ft.

	Tons.
Weight of a 9 ft. cylinder, 30 ft. in height ... =	137·40
Weight of a 7 ft. cylinder, 40 ft. in height ... =	115·92
	253·32
Weight of superstructure and rolling load, as before =	141·00
Total =	394·32

REQUIRED DIAMETER.

SUPPORTING POWER.

	Tons.
Area of a 9 ft. cylinder = 63·62 sq. ft. × 5 tons =	318·10
Surface friction 29·06 × 24 ft. × ⅛ of a ton ... =	87·18
Total	405·28

It will be noticed that support from surface friction is not taken into account for the first 6 ft. in depth of the ground. The depth relied upon for permanent support being 30 — 6 ft. = 24 ft.

The safe frictional resistance per square foot is calculated at

$$\tfrac{1}{8}\text{th of a ton} = \frac{2{,}240}{8} = 280 \text{ lbs.}$$

The load on the concrete hearting per square foot at the commencement of the reducing ring is as follows:—

The weight of the concrete, in tons, per lineal yard of the height of a 7 ft. cylinder is 7·02 tons.

	Tons.
7·02 tons × 13⅓ lineal yards ... ... =	93·60
Superstructure, etc., as before ... ... =	141·00
Total	234·60

	Square feet.
Internal area of a 7 ft. cylinder ... ... =	38·49

	Tons.
Load on concrete per square foot = $\dfrac{234·60}{38·49}$ =	6·09
Add the strain per square foot from the iron rings which equals	0·70
Total load	6·79

The following tables will be found useful in the calculations required in designing a cylinder bridge-pier or in adopting the well system of foundations for bridges, dock-walls, quays, weirs, or other purposes:—

Table A.

No. 1. Internal diameter of cylinder.	No. 2. Internal area of cylinder in square feet.	No. 3. Contents in cubic yards per lineal yard of the height of cylinder.	No. 4. Weight in tons of concrete in Portland cement per lineal yard of the height of cylinder.	No. 5. Weight in tons of brickwork in Portland cement per lineal yard of the height of cylinder.
4 ft. 0 in.	12·56	1·40	2·30	1·89
4 ft. 6 in.	15·90	1·77	2·90	2·39
5 ft. 0 in.	19·64	2·18	3·57	2·94
5 ft. 6 in.	23·76	2·64	4·33	3·56
6 ft. 0 in.	28·28	3·14	5·15	4·24
6 ft. 6 in.	33·19	3·69	6·05	4·98
7 ft. 0 in.	38·49	4·28	7·02	5·78
7 ft. 6 in.	44·18	4·91	8·05	6·63
8 ft. 0 in.	50·27	5·59	9·17	7·55
8 ft. 6 in.	56·75	6·31	10·35	8·52
9 ft. 0 in.	63·62	7·07	11·60	9·54
9 ft. 6 in.	70·89	7·88	12·93	10·64
10 ft. 0 in.	78·54	8·73	14·32	11·79
10 ft. 6 in.	86·59	9·62	15·78	12·99
11 ft. 0 in.	95·04	10·56	17·32	14·26
11 ft. 6 in.	103·87	11·54	18·93	15·58
12 ft. 0 in.	113·10	12·57	20·62	16·97
12 ft. 6 in.	122·72	13·64	22·37	18·41
13 ft. 0 in.	132·74	14·75	24·19	19·91
13 ft. 6 in.	143·41	15·91	26·09	21·48
14 ft. 0 in.	153·94	17·11	28·06	23·10
14 ft. 6 in.	165·13	18·35	30·09	24·77
15 ft. 0 in.	176·72	19·64	32·21	26·51
15 ft. 6 in.	188·69	20·97	34·39	28·31
16 ft. 0 in.	201·06	22·34	36·64	30·16
16 ft. 6 in.	213·83	23·76	38·97	32·08
17 ft. 0 in.	226·98	25·22	41·36	34·05
17 ft. 6 in.	240·53	26·73	43·84	36·09
18 ft. 0 in.	254·47	28·27	46·36	38·16
18 ft. 6 in.	268·81	29·87	48·99	40·33
19 ft. 0 in.	283·53	31·51	51·68	42·54
19 ft. 6 in.	298·65	33·18	54·42	44·79
20 ft. 0 in.	314·16	34·91	57·26	47·13
2 ft. 6 in.	330·07	36·68	60·16	49·52
21 ft. 0 in.	346·36	38·48	63·11	51·95

The weight of Portland cement concrete is taken at 136 lbs. per cubic foot, or 1·64 tons per cubic yard.

The weight of brickwork in Portland cement mortar is taken at 112 lbs. per cubic foot, or 1·35 tons per cubic yard.

Column 3, when multiplied by the height of cylinder, in lineal yards, will give the contents in cubic yards.

Columns 4 and 5, when multiplied by the height, in lineal yards, for which the Portland cement concrete, or the brickwork in Portland cement mortar, extends, will give respectively the weight in tons.

The *internal* areas of the cylinder only are given, as they alone are required in calculating the sustaining power derived from the area of the base of the cylinder, as the weight of the bridge rests upon the hearting and not on the ironwork.

In the tables the internal diameters of the cylinder are commenced at 4 ft. and increase by increments of 6 inches to 21 ft. The former may be considered as nearly the least practical diameter of a cylinder foundation. The cylinder, if of less diameter, would partake more of the nature of a pile or column, being of *itself* the support to the superstructure of the bridge, and not, as is the case in an iron cylinder bridge-pier, merely the skin, as it were, *containing* the hearting which actually supports the weight of the superstructure of the bridge.

For ease of calculation the contents are given in cubic yards per *lineal yard* of the height of cylinder, as brickwork and concrete are usually measured by the cubic yard.

TABLE B.

No. 1. Internal diameter of cylinder.	No. 2. Thickness of cast iron in cylinder ring in inches.	No. 3. Weight of cast iron in cylinder in tons per *lineal foot* of the height of cylinder.	No. 4. Surface area in square feet, in contact with earth, per *lineal foot* of height of cylinder.	No. 5. Loss of weight from immersion in water in tons per *lineal foot* of height of cylinder.
4 ft. 0 in.	¾	·160	12·96	·372
″	1	·215	13·09	·380
″	1¼	·269	13·22	·387
″	1½	·324	13·35	·395
4 ft. 6 in.	¾	·180	14·53	·468
″	1	·241	14·66	·476
″	1¼	·302	14·79	·485
″	1½	·364	14·92	·493
5 ft. 0 in.	1	·265	16·23	·584
″	1¼	·334	16·36	·594
″	1½	·403	16·49	·603
″	1¾	·472	16·62	·613
5 ft. 6 in.	1	·292	17·80	·703
″	1¼	·367	17·93	·713
″	1½	·442	18·06	·723
″	1¾	·517	18·19	·734
6 ft. 0 in.	1	·318	19·37	·832
″	1¼	·400	19·50	·843
″	1½	·481	19·63	·855
″	1¾	·563	19·76	·866
6 ft. 6 in.	1	·344	20·94	·972

TABLE B (*continued*).

No. 1. Internal diameter of cylinder.	No. 2. Thickness of cast iron in cylinder ring in inches.	No. 3. Weight of cast iron in cylinder in tons per *lineal foot* of the height of cylinder.	No. 4. Surface area in square feet, in contact with earth, per *lineal foot* of height of cylinder.	No. 5. Loss of weight from immersion in water in tons per *lineal foot* of height of cylinder.
6 ft. 6 in.	1¼	·432	21·07	·985
,,	1½	·520	21·20	·997
,,	1¾	·609	21·33	1·009
7 ft. 0 in.	1	·370	22·51	1·124
,,	1¼	·465	22·64	1·137
,,	1½	·560	22·77	1·150
,,	1¾	·655	22·90	1·163
7 ft. 6 in.	1	·396	24·086	1·286
,,	1¼	·497	24·216	1·300
,,	1½	·599	24·347	1·314
,,	1¾	·700	24·478	1·328
8 ft. 0 in.	1	·422	25·656	1·459
,,	1¼	·530	25·787	1·474
,,	1½	·638	25·918	1·489
,,	1¾	·746	26·049	1·504
8 ft. 6 in.	1¼	·563	27·358	1·659
,,	1½	·677	27·489	1·675
,,	1¾	·792	27·620	1·691
,,	2	·906	27·751	1·707
9 ft. 0 in.	1¼	·596	28·929	1·855
,,	1½	·716	29·059	1·872
,,	1¾	·837	29·190	1·889
,,	2	·957	29·321	1·906
9 ft. 6 in.	1¼	·628	30·500	2·062
,,	1½	·756	30·631	2·080
,,	1¾	·884	30·762	2·098
,,	2	1·012	30·890	2·116
10 ft. 0 in.	1¼	·662	32·071	2·280
,,	1½	·795	32·202	2·299
,,	1¾	·929	32·333	2·318
,,	2	1·060	32·463	2·336
10 ft. 6 in.	1¼	·694	33·641	2·509
,,	1½	·834	33·772	2·529
,,	1¾	·975	33·903	2·548
,,	2	1·116	34·034	2·568
11 ft. 0 in.	1¼	·726	35·212	2·749
,,	1½	·873	35·343	2·769
,,	1¾	1·021	35·474	2·790
,,	2	1·167	35·605	2·810
11 ft 6 in.	1½	·913	36·914	3·021
,,	1¾	1·067	37·045	3·042
,,	2	1·220	37·176	3·063
,,	2¼	1·376	37·307	3·085
12 ft. 0 in.	1½	·952	38·485	3·284
,,	1¾	1·113	38·615	3·306

Table B (continued).

No. 1. Internal diameter of cylinder.	No. 2. Thickness of cast iron in cylinder ring in inches.	No. 3. Weight of cast iron in cylinder in tons per lineal foot of the height of cylinder.	No. 4. Surface area in square feet, in contact with earth, per lineal foot of height of cylinder.	No. 5. Loss of weight from immersion in water in tons per lineal foot of height of cylinder.
12 ft. 0 in.	2	1·273	38·746	3·328
	2¼	1·434	38·877	3·351
12 ft. 6 in.	1½	·991	40·055	3·557
"	1¾	1·158	40·186	3·580
"	2	1·325	40·317	3·603
"	2¼	1·493	40·448	3·627
13 ft. 0 in.	1½	1·031	41·626	3·841
"	1¾	1·204	41·757	3·865
"	2	1·377	41·888	3·889
"	2¼	1·550	42·019	3·914
13 ft. 6 in.	1½	1·070	43·197	4·137
"	1¾	1·250	43·328	4·162
"	2	1·430	43·459	4·187
"	2¼	1·612	43·590	4·212
14 ft. 0 in.	1½	1·109	44·768	4·443
"	1¾	1·296	44·899	4·470
"	2	1·488	45·030	4·497
"	2¼	1·672	45·161	4·523
14 ft. 6 in.	1½	1·148	46·339	4·760
"	1¾	1·342	46·470	4·787
"	2	1·535	46·600	4·814
"	2¼	1·730	46·732	4·841
15 ft. 0 in.	1½	1·188	47·909	5·088
"	1¾	1·388	48·040	5·116
"	2	1·588	48·171	5·144
"	2¼	1·788	48·302	5·172
15 ft. 6 in.	1½	1·227	49·480	5·427
"	1¾	1·433	49·611	5·456
"	2	1·639	49·742	5·485
"	2¼	1·846	49·873	5·514
16 ft. 0 in.	1½	1·266	51·051	5·778
"	1¾	1·479	51·182	5·808
"	2	1·692	51·313	5·837
"	2¼	1·905	51·444	5·867
16 ft. 6 in.	1½	1·306	52·622	6·139
"	1¾	1·525	52·753	6·170
"	2	1·744	52·881	6·200
"	2¼	1·964	53·015	6·231
17 ft. 0 in.	1½	1·345	54·193	6·511
"	1¾	1·571	54·324	6·542
"	2	1·795	54·454	6·573
"	2¼	2·022	54·585	6·605
"	2½	2·248	54·716	6·637
17 ft. 6 in.	1½	1·384	55·763	6·894
"	1¾	1·616	55·894	6·926

Table B (continued).

No. 1. Internal diameter of cylinder.	No. 2. Thickness of cast iron in cylinder ring in inches.	No. 3. Weight of cast iron in cylinder in tons per *lineal foot* of the height of cylinder.	No. 4. Surface area in square feet, in contact with earth, per *lineal foot* of height of cylinder.	No. 5. Loss of weight from immersion in water in tons per *lineal foot* of height of cylinder.
17 ft. 6 in.	2	1·848	56·025	6·958
,,	2¼	2·081	56·156	6·991
,,	2½	2·313	56·287	7·023
18 ft. 0 in.	1½	1·423	57·334	7·288
,,	1¾	1·662	57·465	7·321
,,	2	1·901	57·596	7·353
,,	2¼	2·141	57·727	7·387
,,	2½	2·380	57·858	7·421
18 ft. 6 in.	1½	1·463	58·905	7·693
,,	1¾	1·708	59·036	7·727
,,	2	1·953	59·167	7·761
,,	2¼	2·200	59·298	7·795
,,	2½	2·446	59·429	7·829
19 ft. 0 in.	1½	1·502	60·476	8·108
,,	1¾	1·754	60·607	8·143
,,	2	2·006	60·738	8·178
,,	2¼	2·259	60·869	8·214
,,	2½	2·512	60·999	8·249
19 ft. 6 in.	1½	1·541	62·047	8·535
,,	1¾	1·800	62·177	8·571
,,	2	2·061	62·308	8·607
,,	2¼	2·322	62·439	8·643
,,	2½	2·578	62·570	8·679
20 ft. 0 in.	1½	1·581	63·617	8·972
,,	1¾	1·846	63·748	9·009
,,	2	2·111	63·879	9·046
,,	2¼	2·376	64·010	9·083
,,	2½	2·641	64·141	9·120
20 ft. 6 in.	1½	1·620	65·188	9·421
,,	1¾	1·891	65·319	9·459
,,	2	2·162	65·450	9·497
,,	2¼	2·433	65·581	9·535
,,	2½	2·704	65·712	9·573
21 ft. 0 in.	1½	1·659	66·759	9·880
,,	1¾	1·937	66·890	9·919
,,	2	2·215	67·021	9·958
,,	2¼	2·493	67·152	9·997
,,	2½	2·771	67·283	10·036

The thicknesses of iron in the table are taken from general practice, and are the least and the greatest thickness of cast iron in the cylinder for the respective diameters; many of the thicknesses of metal are those adopted in existing examples.

REQUIRED DIAMETER.

The weights in Table B are the nett weights of the cast iron rings only, and *no* allowance is made for ribs, lugs, or strengthening brackets, bosses, joint flanges, horizontal and vertical stiffeners, for which 20 to 25 per cent must be *added* to the weights given.

Column 3, when multiplied by the height of the cylinder *in lineal feet*, will give the weight of cast iron rings only in cylinder in tons.

Column 4, when multiplied by the depth *in lineal feet* the cylinder is sunk in the ground, and by the frictional resistance of the ground per square foot of the surface area of the cylinder in decimals of a ton, will give the resistance due to surface friction in tons.

Column 5, when multiplied by the depth of water in feet at the lowest tide or depth, gives the flotation power, or loss of weight from immersion of the cylinder, in tons. NOTE.—In shallow rivers, and where the cylinder is of small diameter, this may be disregarded for all practical purposes.

The weight of cast iron per cubic foot is taken, for ease of calculation, at 448 lbs., which $= 0·20 = \frac{1}{5}$th of a ton. The weight of a cubic foot of fresh water is taken at $0·02786$ of a ton.

The forces governing the stability of cylinder bridge foundations may be thus summed up :—

The supporting power is derived from :—
1. The area of the base, which is as the square of the diameter.
2. The area of the surface in contact with the earth, which varies as the diameter and the depth the cylinder is sunk in the ground.
3. The safe load per square foot on the base, or the bearing support due to the internal sectional area of the cylinder.
4. The safe load on the frictional surface per square foot, or the bearing support due to surface friction.
5. The flotation power, or loss of weight from immersion in water, which varies as the square of the diameter and the depth of water.

The non-supporting power is :—
1. The weight of iron in the cylinder.
2. The weight of the hearting in the cylinder.
3. The weight of the superstructure or load on the cylinder from girder and the rolling load.

NOTE.—The first two items vary as the diameter and height of cylinder.

From the preceding statements it will be gathered that the diameter of the cylinder is regulated :—
1. By the weight superimposed, which varies as the span, width of roadway, load, and number of cylinders of which the pier consists.

2. By its own weight, which varies as its own height and diameter.
3. By the depth it is sunk into the ground.
4. By the resistance from friction of the ground on its surface.
5. By the safe load on the base.
6. By its flotation power, or loss of its weight from immersion in water.

It is evident for the cylinder to be stable that the safe load on the base, *plus* the resistance from friction of the ground on its exterior surface, *plus* the flotation power, *must equal* the weight superimposed, *plus* the weight of the cylinder complete; and may thus be expressed :—

Let S = Safe load on the base of a cylinder.
R = Resistance from friction of ground on the surface of a cylinder.
F = Flotation power or loss of weight of the cylinder from immersion in water.
W = Weight superimposed, including the rolling load.
C = Weight of cylinder complete.

Then for cylinder to be stable—
$(S + R + F)$ must not be less than $(W + C)$.

CHAPTER III.

Load on the Base.

Having calculated in detail the required diameter of a cylinder pier for a railway bridge, the load upon the base will be especially examined; but here it is advisable to name a few points to be considered in deciding upon the width of the openings and the form of the superstructure.

In designing most bridges, the chief object is to determine the number of spans required in a certain length to give the necessary stability and utility at a minimum cost; but the nature of the ground may govern the number of openings, as the safe load upon it may not allow of the most economical spans being adopted, because they would cause too great a weight upon the foundations, its even distribution being considered expedient. Also, if the current of a river is swift, the bed covered with boulders to an unknown depth, the safe-bearing soil believed to be inclined, and the shore on each side firm rock, a single span bridge may be the most economical.

In yielding or alluvial soil any form of arch may be objectionable because of the thrust, and also it probably will be well to consider a continuous girder as prohibited. In Holland, because of the frequency of soft, yielding, alluvial foundations, continuous girders are very rarely used. The design may therefore be limited to some form of girder or truss producing as vertical a strain as possible upon the piers or abutments.

The superstructure can be estimated very closely, provided the method of erection has been duly considered in the design, for in some cases the problem of erection is almost the chief element, and overrules many other considerations; but the cost of the foundations cannot be deduced from any formula, and even the nature of the strata may not be known with certainty.

Consequent upon the increase of dead weight as compared with the moving load, arches or girders of large span are not so affected by rolling and sudden loads and vibration as those of small span. It may so happen that for a large span an arch may be the most economical form by which to bridge an opening, but owing to the difficulty of obtaining immoveable abutments or piers to receive the thrust, it may have to be abandoned, and there may also be objections to it from a local difficulty of erection. If the piers are simply braced iron piles or light columns, girders are used in order that any lateral thrust may be reduced to a minimum. It is very seldom that an arch, whether metal, brick, or stone, or a girder of a bridge fails, but weakness in the foundations, being the cause of settlement, results in deformation and ultimate destruction.

The weight of a pier has also to be considered, for if masonry, brickwork, or concrete, whether encased or not, it may be too great for the foundations to bear unless the base is spread out; for taking the weight of Portland cement brickwork at 112 lbs. per cubic foot, and supposing a pier to be 100 feet in height, the load upon the ground at the surface from its own weight would be 5 tons per square foot, and if Portland cement concrete, about 6 tons. In erecting a girder by rolling out, a pier may be severely strained, and beyond a certain span calculations may show that erection by that method may not be advisable for other reasons, and the cantilever-built-out-from-shore-or-pier, or similar system, may have to be adopted with a comparatively light central girder.

For the piers and abutments of a bridge it may be an advantage to employ material which acts as a monolithic mass, such as Portland cement concrete, and not brickwork or masonry, for the joints are of somewhat uncertain and unequal strength, and particularly for the anchor blocks of suspension bridges.

Having very briefly indicated some matters that affect the founda-

tions, it is apparent that in designing a bridge the nature of the foundations, system of piers or abutments, manner, ease, and rapidity of erection, the strongest and cheapest form, character of the load, purpose of bridge, and the best material to use have to be simultaneously considered in the light of the circumstances of each case, all of which may and probably will, greatly vary.

The load upon the base will now be especially referred to.

First, the importance of ascertaining the nature of the earth, the position of the strata, and the depths at which they occur is evident. Borings have frequently proved unreliable, particularly when they are merely superficial, for then a film or crust may be mistaken for a solid rock bed. In any case of importance they should only be trusted for the place where they are made, and not as indicating the nature or condition of the soil over a considerable area. When pits cannot be sunk, it is desirable that the bore-holes should be frequent.

Excavating pits, using test bars, and driving piles are some of the methods of determining the character of foundations, but care should be taken to ascertain in boring that boulders, or thin strata of hard gravel, are not considered to be solid rock. In sand, mud, or soft clay, they can be made by means of an iron pipe and the water-jet system. Experience has proved that boring with an auger is not so reliable as with a tube, such as is used for artesian wells. In the case of augers when boulders are encountered, further boring is usually arrested in that place, and another bore-hole has to be commenced. Trial pits, where practicable, should be preferred to boring, and they should, if possible, be sunk to a depth below the lowest level of the intended foundations. In testing ground by borings, several should be made, as one hole might encounter a boulder or some hard soil, such as indurated clay, and the latter may adhere to the auger and arrest its progress; the specimen then brought up, being crushed and pressed together, will appear to be firmer than the actual condition of the ground, and will usually indicate rock or hard ground at a *higher* level than it exists. If it should be thought that the nature of the strata or their thicknesses may vary over or near the site, the question has to be considered whether it is advisable to lay dry the foundation in order that its characteristics may be known and unequal subsidence prevented. Irregular ground should be avoided in which to sink cylinders or wells, as it is then difficult to effect vertical sinking.

The area of the base is the principal source upon which the stability of a cylinder foundation depends, as it is generally unalterable. A considerable margin of stability should in all cases be allowed, as from the nature of the calculations exact results cannot be attained. The condition of the earth in each case should be considered, and in works of magnitude it is advisable to make experiments extending as long as practicable, and for at least a month; for it is false economy not to carefully ascertain the character, condition, and other circumstances of a foundation destined to support any part of a structure, a failure of which may result in serious consequences. A continuous surface possesses greater sustaining power than the same area in detached portions, as the adhesion of the sides is not destroyed ; similarly the load that a *tenacious* earth will support upon a small area is somewhat greater than over a large area, because the lateral surfaces are relatively larger in proportion to the area, and, therefore, the effect of cohesion is proportionately greater ; but in loose soils it is not so, for cohesion exists but in name, and the ground around would be upheaved upon an excessive load being superimposed. The weight upon the soil on which narrow walls rest, or whenever it is subject to frequent changes in the direction and amount of pressure, should be less than for foundations which are of considerable continuous extent and depth.

In testing the weight any earth will support, it is not so much the first settlement, provided it is not excessive, that it is desirable to know, but whether after the first settlement it ceases, or the earth, as it were, reacts and rebounds, which it may do in firm ground to the extent of one-eighth to half an inch. If so, the ground is not overloaded, and is only being compressed to firmness, and not crushed.

After ascertaining by experiment the pressure any earth will bear over a given area, the object should be to make the soil neither drier nor wetter than that of its natural state when experimenting, and it should be maintained in that condition. In testing the weight which a soft earth will support, some days should be allowed for the sinking of the test platform, and such subsidence should be known periodically by means of careful levels. A month is not too long for a reliable and complete test, as many soft soils continue to yield. In soft clay soils considerable depression often proceeds for weeks after a load has been applied, but, except in peculiar earths, such settlement will ultimately be imperceptible, and will practically cease. Although it may not be absolutely necessary to experiment when the nature of the ground is well known, wherever stability is of great importance, the cost of a practical experiment being so small, there is no sufficient reason why an actual test of the sustaining power of the soil should not be made in the majority of instances, for there are many earths whose friction and cohesiveness can alone be depended upon for resistance to displacement. In such cases the initial pressure upon the earth should not be much exceeded. The character of the load should be considered, whether it is fixed or moving, and allowance be made when the live load is large as compared with the dead weight, especially in sandy soils. The experiments of Professor Stokes, 1849 ; M. Phillips, 1855 ; M. Renaudot, 1861 ; M. Bresse, 1866 ; and recently of Dr.

Winkler, and others, show that the increase of the intensity of strain consequent upon the dynamic effect of a suddenly-applied moving load may be as much as 33 per cent. more than that of the computed statical pressure.

The normal pressure upon a foundation should be considered in determining the safe load upon the base. It is obvious that if the material is excavated, the initial pressure on the soil is removed. In loose, non-cohesive earths the load may be increased when the depth is considerable, as the soil has been subject to a greater normal pressure due to the we'ght of the soil upon it at any depth, but it is not advisable to consider such increase of bearing power of the soil, unless at any depth it is found that the normal pressure augments the bearing power and makes the earth more dense, which may be approximately ascertained by experiment. In such event the load upon the base can be increased by the weight of the normal pressure removed. Supposing 5 tons per square foot was known to be the safe load upon the surface of the ground, and at any depth it was found that the normal pressure of the soil was 2 tons; $5 + 2 = 7$ tons placed at that depth would equal 5 tons at the surface. In the worst case, when the loose earth is of great depth, and it is certain that it cannot be tapped or disturbed at the depth at which it is decided to place the foundations of a structure, and provided the load is not more than the normal pressure, it is not probable that it will subside or slip, as no additional weight is imposed.

Let D = the depth in feet of a foundation from the surface of the ground.

„ W = the weight of a cubic foot of soil in decimals of a ton.

„ P = the normal pressure on the foundation in tons per square foot.

Then $$P = D \times W.$$

EXAMPLE.—

Let the soil be damp sand; then $W =$, say, 0.055 ton per cubic foot. Let $D = 50$ feet; then $P = 50 \times 0.055 = 2.75$ tons. The normal pressure which is removed is therefore 2.75 tons per square foot. Should the safe load on the base be taken as 5 tons, the excess of pressure above the normal pressure is $5 - 2.75 = 2.25$ tons. It is but necessary to examine the loads put upon screw piles, to see that this weight is considerably less than that which might be safely imposed.

A weight of 5 tons per square foot is generally accepted as the safe load on the blade of a screw pile in firm compact sand, the whole area of the blade being usually considered as support. In this case the normal pressure on the soil is not removed, and the 5 tons pressure is an additional load. Although the shaft of the pile displaces some material, it makes the soil more dense, and therefore heavier, in its immediate vicinity.

For the purpose of comparing the load on the base in a cylinder and a screw pile bridge, let the depth of the foundation be 15 ft., the soil damp, firm, compact sand. The excess of load above the normal pressure on the base per square foot, cylinder pier

$= 5$ tons $- P = 5 - (15 \times 0.055) = 5 - 0.825 = 4.175$ tons.

The excess of load, above the normal pressure, on the base per square foot, screw-pile pier=5 tons, or 20 per cent. more pressure than that of the cylinder pier. The same co-efficient of safe-load on the base for a screw-pile foundation, in which the normal load on the soil is not removed, should not be taken for that of a cylinder pier, because in the former case the foundation is unseen, and it is not absolutely known whether the blade of the screw has remained uninjured in the process of screwing, apart from the question whether the whole area of the blade of a screw-pile should be considered as support; but on the contrary, in a cylinder pier, the foundation is visible; it is known that the whole area of the base is utilised, and that the normal pressure of the soil is removed. An examination, therefore, of past practice shows that the load on the soil from a screw-pile pier is relatively considerably greater than that of a cylinder pier, notwithstanding that in one case everything is known, and in the other, in great measure, is a matter of conjecture. There is no reason why the load on a cylinder pier base should not be somewhat more than that upon a screw-pile foundation; but it is difficult to estimate the actual load that is imposed on the base of a cylinder bridge pier, because of the surface friction, which acts in supporting the load, and therefore reduces the weight on the foundation. Unequal loading of the base may be caused by wind-pressure on a pier, in addition to the normal or initial compressive strain from the weight of the pier and the load; it should therefore be ascertained whether the resultant of the weight of the pier and load on the structure and the side wind-pressure cuts the base so as not to bring a tensile strain on the windward side, and a compressive strain on the leeward side, of any serious amount on the foundation, or the brickwork, masonry, or concrete. If it can be done economically, the initial compressive strain should balance any tensile strain that may arise on the windward side of the pier from the wind-pressure. The effects of the above strains will be transmitted to the foundation, if the latter is placed nearly on the surface, and the load on the base on the leeward side may be considerably increased above the initial strain from the weight of the pier and the load. It is therefore obvious that in such situations the load on the base should be less than if the pressure was always stable, hence the importance, in order to ensure equal pressure on the base, of the filling in a cylinder bridge pier, and the material of any other similar structure, acting as one mass, and being of uniform quality at any level, and of the pier being firmly supported around its circumference or sides by the soil. In the case of foundations of considerable depth, these variable forces will be distributed, provided the pier acts as a monolith, so that their effects will not be materially felt on the base.

In the case of soft strata of great depth, and where it is impracticable to obtain firm, hard foundations, a good plan is to weight a few piles every 4 or 5 ft. or so as they are driven, and to note on each occasion the weight they will bear without sinking; it can thus be ascertained whether the bearing power increases with the depth sunk, and the best depth to place the foundations is known. The experimental piles should be driven in different places, and over the area of the foundation; they must not be drawn, but cut off.

Cylinder and pile foundations should be weighted with a load equal to the greatest they are likely to be required to sustain, and the equally distributed load on the hearting of the cylinder should be allowed to remain for some days, to see if there is any settlement, and the longer this temporary weighting is continued the more reliable is the test. Careful daily observation should be taken to ascertain if any subsidence has taken place; there will almost always be some subsidence under a test load, and it will probably vary in different piers; and although the testing may delay the progress of the works, for the sake of safety it is well that it should be done, unless other opportunities offer of obtaining a true test during sinking operations, and after they are completed, without incurring the expense and delay of temporary loading, which may be a costly undertaking.

It has been proved by practical experience that materials uniform in size and homogeneous in character form the most compact and impenetrable masses. The great stability of breakwaters formed of materials of uniform size, and the firmness of macadamised roads, are proofs of this. The same rule applies to soils. It is the separation of the larger bodies from the smaller that causes a want of cohesiveness and weight-sustaining power.

Near the mouths of rivers, islands often consist of detritus liable to be washed away by a stronger flood than that which deposited them, and also to be eroded by the action of waves, therefore the actual site of a river pier should invariably be examined before sinking a cylinder, although the nature and position of the strata may have been thoroughly ascertained, because stumps or logs of trees or wreckage may be embedded in the soil, which should be removed before commencing operations. Where a stratum of good bearing soil, such as gravel, overlies soft ground of great depth, by increasing the bearing area on the firm earth it may be unnecessary to go to any great depth, but provision must be made against scour.

In foundation and general work, rocks are usually not loaded with a greater weight than from 8 to 18 tons per square foot, according to the character of the rock. As the crushing strength has often been ascertained from cubes, and not from prisms, rectangular blocks, or irregularly-shaped pieces, and as the resistance of rocks to transverse strain or breaking across is considerably less than the compressive strength, and varies greatly, and not always according to the crushing resistance of the material, from 8 to 20 tons per square foot is a prudent limit for the safe load, and should not be exceeded, unless under exceptional circumstances, as unequal bearing may greatly intensify the strain, and irregularity in the texture may reduce the resisting powers to that of the weakest part. Sandstone rock that can be crumbled in the hand should not be loaded with more than $1\frac{1}{2}$ to $1\frac{3}{4}$ ton per square foot, but the strength and weight of sandstone varies considerably. Reference to authorities on the resistance of stones to crushing, tension, and transverse strain, will give the approximate safe load per square foot ; but in foundations, *i.e.*, on the rock in its natural location, it should not exceed one-tenth of the ultimate resistance, and the compressive strength should not alone be taken as a guide to the safe load, but the resistance of the rock to tensional and transverse strain should always be considered in foundation work. The value given for the particular sandstone rock named is for the softest earth that can be called rock, and is merely stated to show that, although some earths may be generally classed as rocks, their bearing power may be limited. The safe load upon an artificial rubble or rock mound foundation depends upon its character, firmness, and solidity when deposited, and upon that of the ground on which it is placed. No general value can be named, although it may be classed as clean or compact gravel.

The following values of the safe direct compressive load per square foot on soils have been carefully compiled from actual practical work, but, of course, are only intended as a guide to the safe load on any earth. The condition of the soil in each particular case must be taken into consideration, and in works of any magnitude, experiments should be made under the same conditions to which the permanent works will be subject, and with the ground both wet and dry. For ordinary conditions of soil, and for the usual depths of foundations, which are assumed to be beyond weather influences, the values given will be found to be approximately correct. The normal pressure or initial weight on the base from the soil is not taken into consideration.

Description of Earth.	Approximate Safe Maximum Load in Tons per Square Foot.
Bog, morass, quicksand, peat moss, marsh land, silt	0 to 0·20
Slake and mud, hard peat turf	0 to 0·25
Soft, wet, pasty, or muddy clays, and marsh clay	0·25 to 0·33
Alluvial deposits of moderate depths in river beds, etc.	0·20 to 0·35

> NOTE.—When the river bed is rocky, and the deposit firm, they may safely support 0·75 ton.

Diluvial clay beds of rivers	0·35 to 1·00
Alluvial earth, loams, and loamy soil (clay and 40 to 70 per cent. of sand), and clay loams (clay and about 30 per cent. of sand)	0·75 to 1·50
Damp clay	1·50 to 2·00
Loose sand in shifting river bed, the safe load increasing with depth	2·50 to 3·00
Upheaved and intermixed beds of different sound clays	3·00
Silty sand of uniform and firm character in a river bed secure from scour, and at depths below 25 ft.	3·50 to 4·00
Solid clay, mixed with very fine sand	4·00

> NOTE.—Equal drainage and condition is especially necessary in the case of clays, as moisture may reduce them from their greatest to their *least* bearing capacity. When found equally and thoroughly mixed with sand and gravel, their supporting power is usually increased. All the values given are for foundations at depths beyond weather influences.

Sound yellow clay, containing only the normal quantity of water	4·00 to 6·00
Solid blue clay, marl and indurated marl, and firm boulder gravel and sand	5·00 to 8·00
Soft chalk, impure and argillaceous	1·00 to 1·50
Hard white chalk	2·50 to 4·00
Ordinary superficial sand beds	2·50 to 4·00
Firm sand in estuaries, bays, etc.	4·50 to 5·00

> NOTE.—The Dutch engineers consider the safe load upon firm, clean sand at 5½ tons per square foot.

Very firm, compact sand, foundations at a considerable depth, not less than 20 ft., and compact, sandy gravel ... ... 6·00 to 7·00

> NOTE.—The sustaining power of sand increases as it approaches a homogeneous, gravelly state.

Firm shale, protected from the weather, and clean gravel 6·00 to 8·00
Compact gravel ... ... 7·00 to 9·00

> NOTE.—The relative bearing powers of gravel may be thus described :—
> 1. Compact gravel.
> 2. Clean gravel.
> 3. Sandy gravel.
> 4. Clayey or loamy gravel.

Sound, clean, homogeneous Thames gravel has been weighted with 14 tons per square foot at a depth of only 3 to 5 feet below the surface, and presented no indication of failure. This gravel was similar to that of a clean pebbly beach.

CHAPTER IV.

SURFACE FRICTION.

THE lateral frictional resistance of the soil on a cylinder pier or well is most frequently not considered as a means of reliable support. The vibration to which most structures are subject tends to destroy surface friction, and the latter is often of an irregular character. Boulders will sometimes hold a cylinder, and if digging out round the base of the column during sinking is adopted, the friction will be reduced. The process of sinking cylinders and wells lessens the surface friction of the soil, because of its loosening the external earth, much of which often gets forced up into the interior of the cylinder, and its place will be supplied by soil having but little cohesion with the firm ground around the cylinder, although the loose earth will become consolidated after a time, and may have of itself considerable frictional resistance. When the earth has been so loosened, or previously dredged, until consolidation has been effected by time and settlement, which usually increases the friction of repose, the surface friction is too variable and uncertain to be safely trusted. Thus, when cylinders or wells are sunk close together in loose, granular soil, the coefficient of surface friction is less than when they are placed at considerable intervals, although ultimately it may be of the same value.

Percolation of water and air lessens the frictional resistance of all soils, but in sinking through most clays, if water reaches them, they will swell and grip a cylinder. This is especially the case when a thin bed of clay occurs in sandy soil. The gripping action will not, however, give a true coefficient of the friction between damp clay and any material, and as the quantity of water in clay in its natural condition varies from about 5 to 13 per cent., the coefficient of surface friction will also not be the same for each kind of clay.

The level of saturation of permeable soil often varies with the water level in a river, and therefore the support from surface friction will change, and may almost be destroyed to a certain depth, as some portion of the earth in contact with the cylinder may be in a state approaching saturation while other parts are nearly dry.

When a hole has to be dredged in the centre of a cylinder to a depth considerably below the cutting edge, in order to cause a cylinder to sink through a stratum of clay resting upon sand, the latter will rush in when the clay bed is perforated, and this action may so disturb the soil around the cylinder as to decrease the natural surface friction for some time after a bridge is finished, for the disturbance of the soil will cause local settlement which will proceed until the earth becomes consolidated. If the surface of iron is neither greased nor lubricated, there is usually very little variation of friction due to changes of temperature, moisture and disturbance being the two chief agents of deterioration, and as the ease with which earths can be disturbed, and their perviousness or imperviousness greatly vary, the coefficients of friction will also be different. Another cause of variability in the frictional resistance is that sometimes a cylinder is not vertical, the friction increasing as it loses its perpendicularity.

Whether the surface friction is uniform in value and immutable are the chief questions to determine.

It is important to remember, should the support which a column may receive from surface friction be suddenly removed by scour, or any of the agencies previously and hereafter mentioned, that an impactive force of serious amount will at once be brought upon the base; the sudden sinking of cylinders after being earthbound proves that the surface friction may quickly cease, or become very small.

No doubt the frictional resistance of some soils is great, and the fact that the stability of timber piles is principally dependent upon it for support, and almost wholly so in soft sand and such soils, shows that surface friction can be trusted in some cases.

If the soil through which a column or pile is driven is of the same character, and there is no probability of the frictional resistance being disturbed, a certain amount of support may be calculated from this source, and although it is impossible to give any absolute coefficient for the friction on the surfaces of piles or columns, a close approximation may be attained by a comparison of soils and circumstances. A crucial test is obviously the most reliable way of ascertaining the frictional resistance, as even the same strata, under different conditions, will give various results, although the values may not deviate to any great extent.

Friction during motion is generally considered to be less than the force necessary to overcome it when at rest, and undoubtedly this is the case when the surfaces are similar, and are smooth and hard and not easily impressed, as iron, granite, concrete, and metals generally ; but when they are comparatively soft and incapable of resisting indentation at any pressure that they may have to bear, the difference between the coefficient of friction during motion and that at the commencement of motion or of repose will not be so marked ; for other resistances may come into action not due *solely* to surface friction of the mass. A surface may become indented or roughened, thus offering opposition to motion not existing at the commencement of movement, and particularly so in any earth of a mixed character possessing hard particles, such as boulders or sand in clay. On the other hand, in the case of hard rock, solid clay, or other homogeneous earth, the difference between friction during motion, and that of friction at rest may be reliably determined. In soils of a granular or gritty nature, small particles become detached during motion, and by pressure occupy or become wedged into any cavities upon the surfaces, and therefore offer resistance which is not *alone* due to friction of a mass upon a mass. From this cause, friction during motion may seemingly even become greater than during rest, but with material consisting of rounded particles that will not wedge, the friction upon a sliding surface may be lessened by reason of the grains revolving.

Friction is the chief cause of stability in granular soils and those readily affected by moisture, which have for practical purposes no immutable cohesion. In few earths are both cohesion and friction of considerable and reliable value, one or the other quickly becoming impaired or destroyed. Deterioration is caused by such various means that each earth must be separately considered, and also the circumstances under which it is placed. The particles of the earth may be dissolved by water and become in a muddy state, or they may be considered insoluble as in clean sand and gravel, although in compact sand or gravel the cementing material may crack and weather. Cohesion may be also more quickly impaired by certain action than friction, and *vice versâ*. It is advisable to ascertain that any earth is uniformly affected throughout the mass, and to prevent or provide against deteriorating influences, for it is useless to declare any'·earth possesses considerable frictional resistance or cohesion when the power can be quickly dissipated by ordinary atmospheric action, and to rely for permanent stability upon such property. In ordinary earths, not rock, it will generally be found that cohesion is small or insignificant in soil having a coefficient of friction of some moment, and the reverse. In most earths friction, although it is affected in a greater degree by vibration, has to be relied upon, and not cohesion, as the latter is variable and may exist almost unimpaired in a lump which, nevertheless, may become detached because of fissures. The coefficients of friction of different earths are also better known than the cohesion; but how easily even friction is impaired may be gathered from the sudden manner in which cylinders will sink after having hung for days by surface friction, or been held by the transitory expansion of clay.

It has been noticed in sandy soils that the surface friction on a cylinder, when sinking operations are not being carried on, and when the material is being raised from the interior, is different; the latter resistance being from 20 to 25 per cent. less than the former, owing to the disturbance resulting from the sand being forced up through the bottom. The frictional resistance may be also lessened by the method of sinking a cylinder, which subject will be examined in subsequent chapters.

In the case of earths partaking of the nature of sand and gravel, which allow of free percolation of water, the permanent friction depends, within certain limits, on the force with which they are pressed together, if not to such an extent as to make them compact and dense, and they will have their frictional resistance increased with the head of water, if the soil is thoroughly waterlogged; but in impermeable soils there will be no practical increase from this source.

As timber is liable to indentation, the friction will increase, to some extent, with time; but in the case of iron or hard surfaces, it will not be augmented after the soil has assumed its normal condition. It is usually found that when the earth is compact and dense there is less lateral pressure, or surface friction, than in the case of loose and incoherent soils. Gritty soils and clay loams have considerably greater frictional resistance than oily, soft clays. Moisture affects the latter earth more than any other soil usually met with.

In sinking cylinders in mud, if desired, the surface friction can be increased by depositing fine sand against the surface of the column, as the particles of the sand will adhere to it to a considerable extent.

The surface friction of iron piles, or cylinders, per unit is considerably less than that of timber piles, on account of the hardness, smoothness, and evenness of the surface of iron, as compared with the roughness and compressibility of wood. Therefore a coefficient for the supporting power from friction on the surface of a wooden pile will not be applicable to that of an iron column. Excepting in the case of mud and silt, the frictional resistance of unplaned cast iron has been ascertained from practical experiments to be about 25 per cent. less than the values for wood. As the outer cylindrical surface of brick wells is generally smoothly plastered or rendered, or has a coating of Portland cement, it may be considered the same as an unplaned superficies of cast iron.

It is obvious that the permanent safe load from frictional resistance of fine soft drift sand should not be taken as equal to that of firm sand, although the former may cause greater resistance to sinking. Friction upon a dry surface is almost invariably greater than that upon a wetted surface, and is so beyond all question upon any plane lubricated with an unguent. The disturbing and enfeebling effect of water may be judged from a careful analysis of many reliable experiments to ascertain the frictional resistance in the case of the same material in a dry and in a wet state on an unplaned surface of cast iron, and on timber piles. It shows the following results :—

That the frictional resistance of an unplaned surface of cast iron on wet sand is about 16 per cent. less than the resistance on the same material when dry. With wooden piles it is about 12 per cent. less, and about 40 per cent. less in sandy clay and gravelly clay soil.

In sandy gravel the resistances are practically the same, whether the soil is wet or dry.

When both materials are in a wet state, sand gives about 20 per cent. more friction than sandy gravel.

The surface friction of masonry and brickwork on dry clay is reduced by from 25 to 30 per cent. when the clay is wet.

Mixed soils, such as clay loams, loams, sandy loams, usually give less surface friction than either the clay or sand of which they are composed when unmixed ; and it may be stated generally that the resistance from surface friction of the ground increases with the smallness and angularity of the particles composing soil of the nature of sand or gravel.

With regard to the question whether the frictional resistance increases in the same soil according to the depth a cylinder is sunk into the ground, it cannot with safety be assumed that it becomes greater, for although many instances have occurred which proved that it does increase, not a few have shown that it does not. What is the reason of this discrepancy ? Broadly, the different condition of the earth in pervious ground, the depth of water, the manner of sinking a cylinder, the state of the surface of the cylinder, whether sinking operations are continuous or intermittent, and variation in the cohesive power of the soil. Theoretically, the friction should vary with the depth and the lateral pressure. The results of tests taken while sinking cylinders or caissons are here alone considered, and they indicate, especially where granular earth is in a state of saturation, that the frictional resistance increases regularly with the depth, but that in dry earths the increase is small. In granular earths the augmentation in the value of surface friction is more marked than in non-granular soil, and in muddy clay and sandy mud, at ordinary depths, such as 30 to 60 ft., the increase is insignificant. In the case of clean sand, it increases; but in gravelly sand and gravel usually very little below a depth of from 10 to 15 ft.

In calculating permanent support from surface friction the total depth a cylinder is sunk in the ground can scarcely be taken, even if the river bed be secured from scour, for the surface friction for the first few feet is small, and seldom in ordinary sand and clay and gravelly sand beds exceeds 40 lbs. per square foot at about 3 ft. in depth, 80 lbs. at 6 or 7 ft., and 120 lbs. at about 10 ft.

It is advisable to allow a smaller coefficient for surface friction in cylinders of small diameter than for those sunk in the same soil of large diameters, because in a cylinder of large diameter the proportion of its circumference to the area of the base is small. On the contrary, where a cylinder is of small diameter, the circumference is nearly equal to the area of the base. For instance, in a cylinder 4 ft. in diameter, the areas of the cylinder and the circumference are equal; whereas in one of, say, 20 ft. in diameter, the area of the base is five times greater than the circumference.

The safe frictional support in the case of a stable or fixed load, may be taken as more than that with rolling loads, which may cause vibration in the cylinders or piles.

The experiments of Mr. Longridge, M.Inst.C.E., in Morecambe Bay, showed that by vibration the bearing power of driven timber piles was reduced to one-fourth or one-fifth of that when subject to a steady non-vibratory load.

The following values are not especially given for the purpose of determining the safe frictional resistance which may be relied upon as permanent support, but they have been carefully deduced, and may be considered as closely approximate. It is well to make experiments with the soil through which the cylinder is to be sunk, when in a loosened condition, in its normal state, and when impregnated with water. The values stated are for soils in the ordinary condition usually met with in sinking cylinders. They are taken from many practical examples, but it is well to repeat that in the same soil the frictional resistances often greatly vary, owing to the amount of moisture in the earth; the roughness, evenness, and smoothness of the face in contact with the soil; the compactness, looseness, or degree of fineness of the strata; and the manner in which the load is applied, whether suddenly or gradually; and the mode of sinking the cylinder. As the girders in a cylinder bridge-pier do not rest upon the iron rings, but upon the hearting of the cylinder, the only direct connection between the casing and the hearting is by means of the horizontal joint flanges of the cylinder. Should the hearting and the rings be unconnected, of course there can be no permanent support from surface friction. The values on unplaned cast iron are for depths not less than 15 ft.

Description of Earth, and material in contact.	Approximate Surface Friction per square ft. in lbs.
Mud and silt, on dry timber sawn piles	100 to 150
,, ,, on clean, unplaned cast iron	50 to 70
Sandy mud, on clean, unplaned cast iron	150 to 250
Muddy clay and viscous mud, on clean, unplaned cast iron	250 to 400

> NOTE.—The frictional resistance generally increases with mud and silt some 25 per cent. between depths of 6 ft. and 20 ft., but after the latter depth it frequently augments but little.

Silty fine sand, liquid when disturbed by water, on unplaned cast iron	250 to 300
Soft clay, on timber sawn piles	160 to 180
Ordinary sand, on unplaned cast iron	300 to 400
Clean river-bed sand and gravel, on unplaned cast iron	400 to 600
Hard compact clay, with a tenacious surface, on unplaned cast iron	900 to 1,000
Ordinary clay beds, on unplaned cast iron	700 to 800
Sharp sand, on clean, timber sawn piles	1,100 to 1,500
Fine soft drift sand, on clean, timber sawn piles	1,500 to 1,700

> NOTE.—In the case of clays, the gripping action upon a cylinder consequent upon their expansion on exposure to moisture or air may make the surface friction appear to be much larger than that caused by ordinary frictional resistance only. The values are for clay containing the normal quantity of water.

CHAPTER V.

SINKING CYLINDERS; GENERAL NOTES.

IN deciding upon the method to be employed in sinking a cylinder, or the means by which the excavation in its interior shall be effected, no prejudice should exist for the absolute use of one system over that of others ; because each method may be useful under certain circumstances, and the shortness of the season during which piers can be erected, may cause the selection of the method of sinking to resolve itself into almost a mere question of which is the speediest. Open air river-pier foundations are to be preferred, but these can generally only be adopted in shallow rivers of depths such as 10 ft. or so, and when the river is free from heavy floods, but if a cylinder can be sunk to a firm and sufficiently watertight stratum, they can be used to any reasonable depth, as the bottom becomes water-sealed and the water can be pumped out.

A reliable comparative table of the cost of sinking cylinders is difficult to attain, on account of the different circumstances and conditions under which they have to be sunk, for even the time required to sink any column cannot be foretold exactly, as an accident, or difficulty with one cylinder may cause considerable delay, and affect the progress of the others. The rate and cost of sinking, depends upon so many things, such as the nature of the soil, whether it is free from boulders or other obstructions, the absence of "blows," the size of the cylinder, its position as regards another column, the method adopted in sinking, the excavating apparatus used, the depth below the water-level and bed of the river, the number of men employed, and whether they are experienced workmen, and upon other contingencies, that it is impossible to lay down any fixed rate. The nature of the strata, the number of cylinders to be erected, the plant at hand, etc., vary greatly, and in many instances the cost of sinking columns of the same diameter in similar soil has disagreed considerably, owing to local conditions. To obtain a firm foundation for a bridge-pier in a soft river bed with a swiftly flowing current is always a more or less arduous undertaking, and the difficulties increase according to the depth below water.

It is most important that the ground should be of the same character over the whole horizontal area of a cylinder, in order that there may be uniformity in the rate of descent, which is to be preferred to irregular and sudden motion, and that sinking may be vertical, as, if it be harder on one side than another, tilting may be expected, and precautions should be taken to support the side on which a tendency to incline occurs.

Some of the many different ways of sinking cylinders may thus be enumerated.

1. The plenum or compressed air method. The vacuum system being considered obsolete.
2. Forcing down the cylinder by weights, and excavating the material in the interior by means of dredgers and excavators.
3. Forcing down the cylinder by weights, and excavating the material in the interior by means of divers.
4. Forcing down by weights, and by dredging the material in the interior, until an impermeable stratum, such as clay, is reached, then by pumping until the cylinder is dry, or water-sealing the bottom of the cylinder by means of cement concrete.
5. The same method of forcing down the cylinder, and by dredging the earth inside, until rock is reached, when the bed is levelled, if necessary by divers, and sufficient cement concrete is deposited to prevent the water issuing up, the remaining water in the column is then pumped out, and the work proceeded with as on dry land.
6. The same method of forcing down and excavating by dredgers, etc., till the intended depth is reached, the bottom is then inspected, and the work done by means of a diving-bell lowered inside the cylinder. This method of sinking was suggested by Mr. E. A. Cowper, M. Council Inst., C.E.
7. By a combination of the compressed air methods with systems 2 to 5.

It is frequently specified that the cylinders are to be sunk so as to leave the bottom dry, and that the concrete is not to be passed through water. Such a stipulation in permeable strata involves either the adoption of the pneumatic method, or that described in No. 6 paragraph. On the Continent, it is almost always specified that the bottom of the foundations shall be examined, and that the concrete hearting shall not be passed through water; this clause necessitates the use in most cases of compressed air to lay the base dry, but the extreme limit at which it can be used is about 120 ft. below water, at a greater depth some other method of sinking must be adopted; and not only is the working time very small at depths over 80 ft. or so below water, but injury and even fatal results to the men follow, and have followed, its adoption. In some situations there may be necessity for such a clause, but to stereotype it is unadvisable, because of the expense of obtaining a dry bottom, although, of course, an examination of unsubmerged ground is always to be preferred to that of submerged earth. If the compressed air system was used where simple dredgers or excavators would suffice, it appears from a comparison of examples that from three to five times more money would be spent than was necessary; but if excavators were used for soils for which they were not designed, or if many large boulders or obstructions are expected to be met with, the compressed air system would be required, provided the depth is not too great, which question will be hereafter referred to; as the excavation can then be carried on as if on dry land.

The size of a cylinder will, to some extent, govern the apparatus to be used. The pneumatic system cannot be conveniently employed with cylinders of less diameter than 5 ft. Care should be taken in devising apparatus that the men are put in such a position that they can freely work. If rapid sinking is of importance, the weight of the cylinder with the kentledge should considerably exceed the friction on the surface of the cylinder, and it may then become a question to determine whether or not it would be advisable to adopt one large and heavy caisson, although it may be more expensive, than several cylinders which would be comparatively light, and would require a longer time to sink them, the object being to cause the downward pressure to be much in excess of any surface friction.

In a clay soil, cylinders can be sunk by being forced down by weights, and by excavating the material inside after it has from time to time been pumped sufficiently dry. The excavation for cylinders of small and ordinary diameter is sometimes done by divers, but it is slow work and of doubtful economy, except in small cylinders, when sand and loose soil overlie an impermeable clay stratum which, on being penetrated a few feet, will water-seal the cylinder, and enable it to be pumped dry enough for excavation to be completed in the open air. Where possible, sinking cylinders by means of weights and excavating by hand, if the water can be removed by baling or easily by pumping, should be adopted, as being the cheaper methods; but if the water is considerable, or fluctuating, dredgers may be preferable, not only on the ground of economy, but also to prevent a run of soil and diminish the quantity of excavation, as the water being in the cylinder will not allow the earth to rush in. The disturbance caused by the ingress of loose soil may be so great as to move the earth for some distance around a cylinder, and militate against vertical sinking, therefore it may be inexpedient to pump out the water and leave the bottom unbalanced by its pressure. It may be necessary to use compressed air where obstacles such as large boulders are met with, or where the ground is difficult or too hard to economically dredge. In deep foundations a combination of the compressed air system and pumping might be used under certain circumstances, for instance, if a watertight stratum was encountered during sinking by the compressed air method; as after it had been reached and been penetrated to a little depth, any water could be pumped out; but, as a rule, it is not necessary to penetrate a watertight stratum very far, for it generally affords a good foundation. Advantage and economy are gained by the disuse of the compressed air system in putting in the lower portion of the hearting.

In a favourable situation, cylinders can be sunk by merely having a hand-pump or two to keep out the water, a ladder placed down the side of the cylinder, a double-purchase crab winch on a platform over the cylinder, in addition to the ordinary staging. Of course, this method can only be used if the pumps are able to keep down the water sufficiently for the men to excavate in the open air, if not, dredging machinery or divers in helmets, must be employed until a watertight stratum is reached, when the water can be pumped out and the work proceeded with. The river bed over the site of the piers should first be levelled by bag and spoon, or other dredger, as it lessens the amount of the excavation in the cylinder, and gives an even surface upon which to pitch the cutting ring.

Where loose soil, such as mud, silt, sand and gravel, overlies an impermeable stratum at a moderate depth, cylinders can be sunk as follows:—By erecting a pile-staging around the site, and bolting together and calking a sufficient height of rings on the platform to reach, when sunk, a little above the water level; or, if a tidal river, above low water level; they are then lowered by a travelling crane working on the staging, the loose top soil is taken out by a dredger, and as the cylinder sinks fresh rings are added until the impermeable soil is reached, when the water is pumped out and the excavation continued.

If a river bed be dry for a certain season, sinking cylinders may thus be conducted. The cutting ring can be conveyed to the site upon a light temporary railway, and be placed in its correct position, the excavation being done by simple digging, lengths of the cylinder being added, and work so carried out until, upon the water-bearing level being reached, it becomes necessary to use dredgers.

With respect to staging, it will be considered in a subsequent chapter.

Great care should be taken at the commencement of sinking operations that a cylinder is perfectly vertical, as both time and money are thereby saved. A simple plan by which it may be known whether a cylinder is sinking vertically, is by hanging several plumb-bobs outside a column. In a rapid river there is especial difficulty in sinking it truly vertical, and allowance should be made for unavoidable divergence in sinking. It is easier to guide a vertical-sided cylinder or caisson than a trumpet or bell-shaped one, and the former will sink straighter than the latter, and has the best chance of retaining perpendicularity. In ordinary firm soil, when cylinders are carefully guided for the first 10 to 15 ft. of sinking, and to 20 to 30 ft. in loose soil, provided the rings are tightly and truly bolted together, they generally go down vertically. To make the sinking uniform, and to prevent tilting, the excavation should be effected on every side as equally as possible.

Perhaps the best way to proceed is first to excavate in the centre, and then to work from there towards the circumference in all directions. The levels of the excavation in the cylinder should be constantly ascertained, so as to keep the bottom as nearly level as practicable.

It is desirable to know if there is a probability of the cylinder sinking suddenly many feet. In that event it may destroy the staging, and perhaps not sink vertically. A thorough knowledge of the strata will generally enable this point to be decided, but in sinking the first cylinder there should be special precautions against such an occurrence, and the manner of its sinking should be noted as a guide to the probable penetration of the other columns.

As both cylinders and wells have become inclined from having the material scooped out under the cutting-ring to a considerable depth when hanging from surface-friction only, and from this friction being suddenly overcome, it is advisable not to allow the column to be suspended from surface friction more than three to four feet, according to the diameter, above the bottom of the excavation in the cylinder During the commencement of sinking operations the column should be most carefully watched, to see that it is sinking equally, and that there is no tendency to incline in one direction.

If cylinders have to be sunk near buildings, and through sandy or loose soil, precautions should be taken against the internal excavation in the cylinder disturbing the foundations of structures in the vicinity, and therefore as little material as possible should be removed. If any buildings or wells near the site show signs of cracking, the excavations should be at once stopped to see what further preventive measures are necessary.

Experience in several cases has shown that when two cylinders have to be sunk close together, or where the distance between them is not greater than the diameter of the cylinder, they should be sunk alternately, as there is a tendency, when they are being sunk simultaneously, to draw towards each other. There is always considerable difficulty in sinking cylinders close together in sand and loose soil, and as a rule it is easier to sink one large cylinder than two smaller ones in making a pier. In some instances the tendency to draw together has been counteracted by having one cylinder sunk half a diameter in advance of the other. Should two parallel rows have to be sunk very near to one another, say 2 or 3 feet apart, one row should be sunk before the other, or they can be started at different ends, or from the centre towards the ends, the object being to disturb as small an area as possible of the soil in the locality of the cylinders at any one time. It is also advisable to sink the columns that are in one line alternately in preference to sinking the next adjacent. The reason of this tendency to draw is believed to be that the sand or soil around the column is in an agitated state, owing to the sinking operations, and if two cylinders are sunk close together the soil between them will be the softest and most loose, and there will therefore be a tendency to cant over at that point. Excavating in a cylinder in sandy soil sometimes throws the neighbouring columns out of the perpendicular when they are sunk in close proximity, and they are also liable to become jammed. Particularly when cylinders have to be sunk to considerable depths, the interval between them should be as much as possible, in order to prevent contact caused by deviation from perpendicularity in sinking.

In sinking a cylinder to a hard stratum which dips at a considerable angle, it may fall over when being sunk. This tendency can be provided against by supporting it by tackle at two or three places in its height. A cylinder must also be secured where the soil is of a soft, semi-fluid character, or of a rocky nature. In sinking them in a silty bed where there is considerable range in the tides, special means should be taken to prevent overturning, as when the tide rises the weight of the column is reduced considerably if the water is excluded, and the effects of the current are more severely felt. In such situations the cylinder should be sunk as rapidly as possible, so as to obtain a good hold in the ground to counteract the tilting force. In sinking cylinders in tidal waters having a great rise of tide, the column must be prevented from floating at high water, because, if it excluded the tide, particularly at the commencement of sinking operations, it might be lifted, provided the bottom was closed, after work was suspended, also before inflating the air chamber, when using the compressed air system, it is necessary to know that the cylinder at all times is sufficiently heavy to obviate floating. If it should be necessary at any time to flood a cylinder or caisson in using the same method of sinking, the object should be to substitute the compressed air by the water, so as always to maintain the same pressure; for if the air be discharged before the water reaches the roof there will most probably be a sudden sinking of the cylinder. The escape of the air can be regulated by the pressure gauges, the air pumps being worked or stopped according as pressure is or is not wanted.

A method of testing whether cylinders have reached firm ground without the aid of divers, or requiring the bottom to be made dry, is by having several borings made around them and near to them ; but this system is not always reliable, and can only safely be adopted where there is no doubt about the nature, thickness, and position of the different strata; and it is but a makeshift, for divers should be sent down to clear away any rubbish that may have accumulated at the bottom, and to level the base for the hearting, which should always extend closely around the cylinder, should it be decided that it is unnecessary to make the foundation dry before depositing the concrete in a cylinder when sunk to the intended depth.

It is important, particularly in sandy soils, to prevent the ingress of the earth into a cylinder. To obviate or lessen such an occurrence, the column should be sunk at a rate corresponding to that at which the excavation is removed, but at first sinking should be slow, until the cylinder has taken a fair bearing, when it should be gradually increased.

Outside scour or subsidence of a river bed during sinking operations, with the consequent rush of soil into the cylinder, is sometimes checked by bags filled with clay or impermeable earth being deposited round the outside of the column. It is not advisable to use stone or a hard substance for this purpose during sinking, although a most excellent material to prevent scour when the cylinder is sunk to the required depth, because the stones may get under the cutting edge, and then will impede easy and vertical sinking. Loose stones in any soil are generally troublesome in cylinder sinking, and are to be regarded as obstructions. A system sometimes adopted is to tip, before insertion of the cutting ring and during sinking operations, clay on and around the site where the cylinders are to be sunk, so as to lessen the ingress of water while pumping. As sandy soil is the most frequent earth to "blow" and rush in, it may be well to remember that perfectly clean sand seldom becomes quicksand, but that a small admixture of clayey matter is sufficient to enable it to be in a condition ready to be converted into a quicksand.

As a rule, in tidal waters, the greatest downward motion of a cylinder may be anticipated at low water. In driving or sinking hollow piles or columns in quicksand, the sand will run in according to the depth of the bed and outside head of water. Should the water be pumped out from the interior, weighting will often stop this ingression. During sinking operations, if the compressed-air system is not used, the cylinder should be kept full of water in order to prevent a run of sand at the base. When the water is lowered in the cylinder in sandy and loose soils, a "blow" will frequently occur, that is, the soil will rush up from below, and in very loose material may nearly fill the cylinder. "Blows" will also occur in cylinder sinking from the compressed air rushing out under the cutting edge. They may then be arrested by slightly lowering the air pressure, which must be done very carefully, or the water may come in and endanger the lives of the men. A "blow" may be permitted to continue for a short time if it is found that a return wave, as it were, of water takes place. "Blows" are mainly caused in loose porous soil by changes in the water level of the river, whether owing to great range of tide, or wave action. When a run of sand takes place in a cylinder, and it has been stopped by letting in water, or by a layer of stone, gravel, clay, or other means in the cylinder, before recommencing sinking operations the sand should be allowed time to subside and get into a state of rest. On the other hand, when no "blows" occur, in mixed soils, or where bubbles of air appear through the water, sinking should be carried on without intermission, so as to prevent the soil settling and subsiding, and the smaller particles incorporating with the larger and becoming consolidated.

To prevent a temporary rush of sand or loose soil from entering a cylinder, to lessen the disturbance of material round it, and to give it time to settle, a moveable diaphragm can be placed upon the bottom ring of the cylinder, with a valve opening inwards, which if shut stops the ingress of the material. It is important that the inside of a cylinder should not become choked by the ingression of earth, as then not only is there more soil to excavate, but the weight required to sink it will be increased, for there will be the interior surface friction to overcome as well as the exterior, and two simultaneous resistances to penetration instead of one; therefore the internal sides should be kept free from contact with the earth in the cylinder. Again, the soil that is forced into the cylinder frequently only comes from one side, and that the softest and loosest, then sinking will probably not proceed vertically and the column may be drawn towards the side where the soft or loose earth occurs. In sinking cylinders through quicksand or very soft soil, the bed of the river outside the column should be watched to see if its surface subsides; should this be the case, it shows that there is a run of the soil. In most instances the quantity of material taken out of the cylinders in loose soil is greater than the contents of the column. In cohesive soils the increase may be but little, perhaps not more than 20 per cent.; but in deep beds of sandy and loose earth, it is seldom less than from 40 to 100 per cent. in excess of the contents of the subterranean portion of the cylinder. As this excess of soil must come from the outside of a cylinder and be drawn in, it disturbs and cracks the surrounding earth and contributes to prevent vertical sinking. Should "blows" occur in a quicksand, the cylinder may become filled with sand and water to the water-level outside. The external and internal pressures should be balanced, or but a slight preponderance of water outside should be allowed to obviate any tendency of loose soil to "blow," and the water inside should be lowered or raised as the outside water ebbs or flows, for the excavating apparatus may take out of the cylinder, with the earth, more water than percolates in the same time. If it is possible to balance the waters during sinking, comparatively little more material than the contents of the cylinder will require to be removed, and "blows" will be prevented. Should the sinking be suddenly arrested, it may often be resumed by lowering either the air-pressure or the water inside the cylinder, the usual result being that the water then infiltrates through the earth outside into the interior of the cylinder, and consequently loosens the soil upon which the cutting ring rests. It will generally occur that when it is low water, and a column of water is in the cylinder considerably higher than the level of the river outside, less earth will be brought up by dredgers than when the water-levels nearly correspond. Where the dredging system of sinking is adopted, the best plan is to make the depth of water inside identical, or nearly so with that outside, and to rely upon deadweight to cause the cylinder to penetrate ; but when the soil is open and allows of free percolation it is usually found if the water outside is sufficiently higher than that inside as to cause disturbance of the frictional surface, consequent upon the unbalanced pressure, that cylinders sink easier. In soft and muddy soils the water levels should nearly correspond, any difference increasing as the earth becomes firmer, less pervious, and the particles of which it is composed, harder. In impermeable soil, such as clay, it is advantageous to have the water considerably lower inside than outside, so as to facilitate sinking, as no "blows" are probable. However, if the water outside be allowed to have too great a preponderance, an upheaval of the earth inside will take place, which only increases the quantity to be excavated and jams the cylinder. Therefore, when there is but little earth inside, it is advisable not to draw off any more water as soon as the column begins to sink. Should a river-bed be dry at low water, or the water then be of little depth, excavating in the open may be possible for a few hours each day, and in order to prevent a "blow," water can be admitted into the cylinder as the tide rises outside, and the greatest head of water can be ascertained which the earth can bear without its percolating in such quantity as to stop excavation by hand. It is essential that cylinders be sunk rapidly in sandy and loose soils. An excess of weight for sinking is therefore an advantage, because any little expense in temporarily loading the cylinder is soon compensated ; as, by reason of the rapid sinking, the outside soil will generally be prevented from entering the cylinder to any great extent. With the compressed air system among some of the expedients used to lessen the ingress of the soil into the cylinder, it has been found that by working the air-pumps rapidly, and by sending in the maximum quantity of compressed air some time before operations are commenced upon each day or shift, and during work, and by enlarging the base of the cylinder a little beyond the diameter of the bottom ring, the water is driven to some extent out of the sand or loose soil which consequently possesses more cohesion, and does not run so quickly into the cylinder. It may also be advisable to excavate at opposite points, but in doing this the column must not be tilted. It has been suggested that where the soil is of an extremely moveable character, the material should be expelled outwards and not taken through the cylinder.

In sinking cylinders through sand with a thin bed of clay intervening, the columns will very probably become earth-bound, and although they may sink through the clay into the sand, if water percolates to some clays they swell, and will grip the cylinder, but by excavating the sand so as to cut away the stratum of clay touching the cylinder it will go down at once. When cylinders hang or refuse to sink, notwithstanding the earth being excavated from beneath, on the water being lowered in the cylinder some 6 to 8 ft. below the outside water level, they will often at once go down, the pressure on the bottom being reduced, and a flow created at the base. With brick wells this is a somewhat risky operation, and the better way is to keep the water at nearly the same level, rather lower on the inside, in order to obtain pressure from the outside, but the nature of the soil, and special circumstances must decide this point. As in sand and similar soil, the more it is agitated on the surface of the cylinder the less the frictional resistance becomes, small perforated pipes can be put down round the outer circumference of the cylinder, and on water being pumped or discharged into them under considerable hydrostatic head, the surface becomes disturbed. This system has been used with success and economy to sink wooden piles in sand, into which they could only be driven with great difficulty.

It is possible, as large tarpaulins spread over the bottom of a river have been used for passing a breach in tunnel-work, that they might be useful to abate surface disturbance in loose soils in cylinder sinking. They are also of assistance when laid on the bottom and round the sides of a foundation, in preventing cement being washed out of concrete, and in conducting water to a pumping sump, if a considerable flow is encountered.

If the water which comes up from the bottom of the column brings with it mud and sand, and the cylinder will not go down to the proper depth on account of the strata, or some obstruction which cannot be removed by dredging, divers must be employed, or the compressed-air system. In countries, as in India, where at certain seasons river-works cannot be executed, only those cylinders should be commenced that can be completed in one season, or made secure before the floods, and work during their continuance will have to be different from that at other times. In firm clay the cylinder need not penetrate more than from 7 ft. to 10 ft., if protected from scour and movement of every kind; and similarly 4 ft. to 5 ft. in compact gravel, and 3 ft. in rock, where the strata are all of considerable thickness, the depth becoming greater according as the earth is more affected by moisture and other deleterious influences.

In clay, muddy, peaty, or vegetable soil, men often complain of the fetid atmosphere of a cylinder, and instances have occurred in sinking, in which fatal effects have arisen from the noxious gases generated in decayed earth or soil saturated with impure water. It may therefore be necessary to introduce fresh air into a cylinder, and to use disinfectants in order to purify it.

CHAPTER VI.
Sinking Cylinders; Staging; Floating Out.

CYLINDERS are usually properly adjusted and guided in sinking by a temporary timber framework ; but in an exposed situation, and where the bed of the river is treacherous, and there is a great depth of water, fixed staging for erecting cylinders may be economically impracticable or objectionable, and the pontoon system of floating out and sinking may be preferably adopted. High and low level staging is sometimes used, the rings being placed upon the low level stage, and the crane or lifting apparatus on the high level platform. A few guide-piles driven by means of barges or suitable floats, so as to secure the cylinder in a correct and vertical position, and aid its perpendicular subsidence are most useful, or an adjustable frame fixed on the river bottom, the size of the cylinder, is sometimes employed. If the depth is too great for timber guide piles, iron piles can be used.

There is no doubt that a cylinder is far easier kept in a vertical position during sinking than straightened after it is inclined, and a little expense in staging will often save both time and money. In addition to guiding a cylinder, the staging must act as a platform for lowering the rings, the hearting, mixing concrete, and be of sufficient extent so that dredging or excavating, raising and lowering, or compressed air machinery can be placed and easily worked thereon, should it be necessary. In order to lessen the height of the staging, triangular-shaped pockets have been cast on the outside face at the base of the reducing ring at intervals around it into the horizontal seats of which guide poles fitted, and the segments above the reducing ring were fixed from this temporarily fixed staging, no driven piles being required. This system was adopted at the Albert Bridge, Chelsea, England.

As considerable time is occupied in fixing and removing staging, in rivers subject to a prolonged flood season when operations on river piers have to be discontinued, it is necessary that the work be so arranged that there is no occasion to discharge men who may be difficult to re-engage, and in order to economise time it is advisable to require as little staging as possible, and so designed as to be quickly erected and removed. A floating timber gangway, sufficiently wide for conveyance of the materials for the piers, placed across a non-navigable river has been found to much facilitate the work.

It may be well to name a few recent methods of erecting staging and cofferdams for river piers under somewhat exceptional circumstances. Where a streak of rocky soil unexpectedly occurred in erecting the staging and a cofferdam for a bridge-pier, the following expedient was adopted. Iron rods 2 to 2¼ in. in diameter were fixed in the bottom of the timber piles, the lower end of the rods were split in order to receive a fine wedge. Holes were bored in the rock, into which the split rods were dropped, and on their being driven down the wedges caused the ends of the rods to splay out and thus firmly hold in position the foot of the pile. To accelerate the erection of bridge-pier cofferdams, corrugated iron sheets, stayed with timber framing and weighted with rails, have been recently employed in sandy and gravelly soil with success. Rails have also been placed in holes in rock at intervals of from 6 to 10 feet, and walings fastened to the rails, and sheet piles driven between them, puddle being deposited at the base. Corrugated iron plates with overlapping edges which fitted into guide plates that bound the two together, were also used at the New Tay Bridge instead of a timber cofferdam, because of the shallowness of the soft stratum, 6 in. in thickness, above the rock.

Old boats lashed together, and connected by timbers, have also been filled with stone or clay, and sunk and surrounded with material until they were firmly embedded in the river bed and a rigid base formed for the piles of the staging or raking struts. This method of making a foundation for piles and temporarily fixed stage for cylinders has been used when the river bed was rock, and also when it was loose sand.

With regard to the pontoon system of floating out cylinders or caissons for bridge piers, two pontoons or barges with a platform on them are usually employed, with an opening a few feet larger than the diameter of the cylinder in the covering between them, and both pontoons are planked over and firmly connected. In order to lessen the expense of specially built pontoons they are sometimes designed so as to be afterwards used as landing stages, or as river-service vessels.

In a rapidly flowing river of considerable depth it is necessary in floating out to have command over the bottom of the cylinder or caisson as well as the top, as neither can be sufficiently controlled from the top only. Chain cables $1\frac{1}{4}$ in. to $1\frac{3}{4}$ in. in diameter, about 45 to 60 fathoms in length, attached to anchors weighing from 21 to 40 cwts., are frequently used for ordinary cross moorings. Large pontoons require in a soft bed anchors weighing about 3 tons each. To keep the pontoons steady in a strong current the moorings must be strained tightly; a capstan is therefore generally fixed on the pontoon, and a chain housed round a bollard on it. In some of the modern pontoons for floating out cylinders, the chain cables pass in at the bottom of the pontoon, through an inclined cable-trunk over a deck-roller at the top, and are held by a link-stopper attached to an adjusting or straining screw, which works in a gun-metal nut by means of gearing, the whole being secured to the pontoon deck by plate-iron framework. The moorings being accurately laid, the tightening of the chain-cables by the screw-gear adjusts the pontoons exactly on the centre line of the bridge. In all loose soils, experience has proved that it is the scope given to the moorings that makes them secure; therefore, they should never have a vertical hold, although the more a cable is payed out, the more difficult it is to lift the anchor. Should the anchors at first let down not give a good hold, they should be weighed and cast afresh, with a greater length of cable. In quicksand the moorings become faster by time ; but if they should tend to approach the vertical, they should be lifted and recast. A scope or length of cable found in loose soil to give stability to a pontoon, is from twenty to twenty-five times that of the range of the tide ; and, should there be no tide, from ten to fifteen times the depth of water. The more cohesion the soil forming the bed of the river possesses, the steeper can be the inclination of the moorings. Mushroom anchors are found to be most effectual for mooring lightships in deep water. In mooring on an exposed coast, the chains on the offshore side, and against the set of the current, should be stronger than those on the inshore side.

A great improvement on the pontoon system, moored by chain cables, was that adopted for the pontoons at the New Tay Bridge, which were fitted with four large hollow wrought iron cylindrical legs. By means of hydraulic apparatus, these legs could be raised or lowered when the pontoons were floating, and when the legs were firmly bedded in the sand of the river bed, the whole pontoon could be raised or lowered by the hydraulic apparatus. The pontoon had two rectangular openings, large enough to let the bases of the cylinders pass through. On the pontoon there were steam engines for working the excavating apparatus, and a concrete-mixing machine, besides steam cranes for lifting and lowering, and various other apparatus. The caisson was kept perpendicular by the rigidity of the pontoon and without mooring chains.

A temporary bottom is sometimes attached to a cylinder against the internal flanges, so that it floats until it is known that the ground is sufficiently solid to bear its weight. In swift currents difficulty is experienced in floating out cylinders of small diameter, such as from 5 to about 8 feet. If the weather should be stormy it will be economically impossible to steady such cylinders sufficiently for sinking. If desired, pontoons can be flooded so as to partly sink when the column is suspended by them over the site.

The height of the cylinders to be floated out by barges or pontoons depends upon the depth of the water at high tide, and also on the contour and nature of the ground. It is prudent to ascertain whether the level of the bed of the river varies much along the course of towage, and care must be taken that the cylinder or caisson does not ground at low tide. Built-up cylinders from 40 to 60 ft. in height have been floated out and sunk into position before lengths were added to keep them above high-water level. While the cylinder is being lowered from the pontoon, and when floating over its site, two or three lengths can be added. The column should not be left until it has a bearing in the ground of not less than about ¼ of its total height, or in the case of a caisson about ⅕ of its greatest horizontal length, or it may be overturned or floated. In all cases when the centre of gravity of the column is high, precautions must be specially taken against overturning by lashing its top and bottom to the pontoons, or otherwise supporting it. If the compressed-air system is to be used in sinking, by having a movable closed top on the cylinder, air can be pumped in, and the column will float, it can then be carefully brought to the required position, and the air being let out slowly the cylinder will sink. This method is sometimes adopted with caissons built on shore, and launched down an ordinary slipway. Two or three of the bottom rings of a cylinder reaching above low-water mark have been built up on the foreshore of a river, and fixed to two girders resting on temporary wooden supports. At low water barges or pontoons are introduced under the girders, the whole being lifted bodily as the tide rises, with the exception of the supports, which may not always be needed. The connections between the pontoon girders and the cylinder rings should be so made that they can be readily unshipped. Rings of the cylinder or caisson and part of the hearting can be added as the work proceeds, or in a large caisson having buoyancy compartments, concrete can be deposited to help the sinking operations.

CHAPTER VII.

Removing Obstructions in Sinking and "Righting" Cylinders.

ALTHOUGH cylinders can penetrate obstacles, such as logs of timber, sunken vessels, and quicksands, which are difficult to overcome with other systems of construction, still boulders and stony and other débris often give serious trouble, and cause considerable loss of time and money. In some cases there may only be difficulty for the first fifteen or twenty feet of sinking in loose soil, such as silt, mud, and sand, it being then overcome by the weight of the cylinder; on the other hand, the difficulty of removing them may increase with the depth, and this is more probable. Boulders, which often vary in size from that of large stones to masses of rock weighing as much as from six to eight tons; lumps of hard clay and sunken trees are frequently encountered in the holes and depressions in river beds, and often tilt cylinders during the operation of sinking. If the compressed air system is not used, and boulders, trunks of trees, and driftwood, etc., are expected to have to be excavated, divers should be sent down the cylinder at intervals to examine the interior, in order that necessary precautions may be taken. Owing to variation in the size and hardness of boulders in some sandy and gravelly soils, one cylinder may take much longer to sink than another. It is difficult to remove boulders or débris and prevent delays, or to level rock foundations without the employment of pneumatic apparatus, or skilled, not ordinary, divers.

When large boulders are under a cutting edge, it is not an easy matter to remove them safely, for if they are pulled into a cylinder, a "blow" of soil frequently follows, and breaking them up may be a very slow process. The simplest way of removing them, or tree stems, is by pushing them out, by cutting out, which is a somewhat slow operation, or by drawing them into the cylinder, but the latter method cannot always be effected, and in loose soil will probably cause "blows," nor can they always be thrust out. They can also be displaced by splitting up with plugs, wedges, jumpers, by drilling, or by undermining and drawing into the cylinder, and then by breaking them to convenient sizes for raising and discharging. Tree stems and logs have also been cut through under water by means of an axe blade attached to the monkey of a pile driver, and by divers augering out and cutting them to pieces. In situations where, at the commencement of sinking, large boulders have been drawn towards, or have rolled against the cylinder, they have been removed by the column being lifted, and by drawing them within the circumference of the cylinder.

When the compressed air system is used, as the ground at the bottom of the cylinder is not submerged, it is at once seen when boulders or débris are in the soil, and many methods of removing obstructions can be prosecuted with facility, which it would be impossible to employ with effect if water were in the cylinder; and in consequence of the slow rate of progress by other means than visible excavation, the pneumatic method of sinking may become indispensable.

When the boulders extend considerably under the cutting edge, say from two to three feet, circumstances will show whether it is better to pull them into the cylinder, or to chip them to pieces. The latter should be done for a little distance beyond the cylinder, so as to prevent the edges holding the column. To push them out will probably be impracticable. Where boulders are packed together closely with decomposed rock, steel-pointed picks may not excavate or break them up, but heavy steel bars driven in with sledge hammers, or dropped inside a pipe from a considerable height, or guided by other means, may shatter them. If concreted gravel or clay is found between boulders, they will generally not be so difficult to break up as when united with decomposed rock. If it is decided to draw the boulders into a cylinder, as much of the ground as possible in which they are embedded should be first cut away, and in order to lessen any ingress of soil, and to give it time to settle, it may be advisable to temporarily weight the earth at the base of the cylinder during the operation of pulling in the boulders, and for some time after, depending upon the degree of looseness of the ground. The cutting edge of the bottom ring should be protected against damage in attempting to thrust it through boulders, but it is not easy to strengthen it, and the result may be that, if it is of cast iron, it will be cracked and shattered ; if of wrought iron, it will be bent, deformed, and crushed. In all cases it is far the better plan to dislodge the boulders, than to attempt to thrust the cylinder through them, which will usually be found ineffectual, and generally a dangerous operation.

Cylinders are sometimes firmly held by a clay stratum, which is, as a rule, an easier obstruction to overcome than boulders, loose stones, or tree logs ; it may then be advisable to excavate or dredge a hole considerably below the cutting edge in order to cause the cylinder to sink, but as much provision as possible should be made against a run of loose soil into the cylinder. The resistance to be destroyed in such a case is not merely the surface friction, but the gripping action caused by the swelling of the clay. In addition to that named, more weighting ; the drilling of numerous holes in the cylinder through which compressed air can be discharged ; driving hollow perforated iron tubes at intervals on the outside of the cylinder, and ejecting water under considerable pressure through them to soften the clay and reduce surface friction, holes being first made in the clay by augers or boring tools can be tried.

The following expedients might be employed as a kind of last resource. As the swelling of clay is chiefly caused by water, if, without injury, the surface of the cylinder could be made sufficiently hot, the clay would shrink and the surface friction would be reduced. The expansion of the iron rings might also very slightly compress the soil, and on the metal cooling the contraction would probably cause a void, however small, between the outer surface of the cylinder and the earth, independently of the shrinking of the clay should that appreciably occur. As the action of an acid on clay tends to soften it, an acid fluid might be discharged into the clay at intervals, through holes in the cylinder rings, although its effect on the metal casing would be deleterious.

With regard to the employment of explosives for removing obstructions, they have been successfully used in large caissons, the reasonable precautions in blasting in a confined space being observed, but in small cylinders the result generally is that the cutting ring becomes cracked and blown out when of cast iron, and deformed when of wrought iron, but if the ring can be so strengthened as to continue to equally and properly sink, cracks are not of much importance, provided surface friction of the rings against the earth is *not* relied upon for permanent support, as the weight of the superstructure will rest upon the hearting only, but should the cutting ring be so damaged by the explosions as to be fractured, it may become necessary to entirely remove it, which must always be a work of difficulty ; and the absence of the cutting ring in sinking would cause that operation to be conducted with no small amount of danger to the cylinder should more obstructions be encountered. Small charges of from one to two ounces of dynamite have been used for aiding sinking operations, as well as for cutting away rock or boulders that projected into a cylinder, the idea being to overcome friction by tremor and vibration, and to cause the cylinder to sink with a diminished load ; but in other than large caissons the use of explosive agents is questionable, and the drilling of holes by divers is always very tedious work. The hydraulic method, by means of water discharged at considerable pressure, through perforated pipes on the surface of the cylinder, is to be preferred in the case of moderate-sized cylinders, or in a caisson, as it cannot injure the rings. To lessen the chance of the cutting ring being damaged, broken, or blown out, the charge has been placed in a pit excavated in the centre of the cylinder and there exploded; no deleterious effect follows, but a trembling motion is imparted to the earth, the surface friction is reduced, and the cylinder at once sinks. Only very small charges should be used, the charge being increased according to the area of the base, it being borne in mind that *continued* disturbance and easy regular sinking is to be desired, and not intermittent, sudden, and violent penetration, which will be difficult to control sufficiently to cause the cylinder to proceed in a perpendicular direction, and may crush the cutting ring. When the necessary apparatus for the adoption of the compressed air system is not available, and the dredger machinery cannot remove the boulders, other specially-devised apparatus, such as steel bars dropping from a height in guides, can be tried ; but blasting may be the only agency by which they can be shattered. As there are about two hundred different explosive agents, the selection of the most fitted for the work required to be done is best determined by an expert.

With regard to "righting" an inclined cylinder, a generally effective way of getting it back to verticalness is to firmly wedge up the lower edges of the column on the depressed side, and excavate the soil under the uplifted portion. When this has been effected the compressed air, should that system be used, can be suddenly discharged, the consequence being that the material around the bottom of the cylinder will enter into it. The top of the column can also be propped up on the lower side. In some cases, for general sinking purposes, and in order to reduce the surface friction, numerous holes have been drilled through the rings on the higher side of the cylinder; the compressed air escaping through them loosened the material outside and lessened the frictional resistance.

In an instance where a cylinder was sunk 30 ft. in sand, and was considerably inclined, the simple wedge plan did not succeed. The upper edge had to be under-excavated, so that the escaping air passed through and loosened the material on that side. The cylinder was also wedged up on its lower side, and a battering-ram, made of a whole oak balk suspended from shear-legs, was used to strike successive blows on the top of the column on its higher side. During descent it was brought into a vertical position. It is preferable to weight the cylinder on the higher side, instead of driving it, but the weights must be arranged so that they can be readily removed. If the segments are firmly bolted together, and the cylinder has become inclined in loose soil before many lengths of rings have been sunk, pulling it over by means of cables and blocks and crabs may be tried in conjunction with any of the other methods herein described ; but there may not be sufficient purchase or hold if the river-bed is mud or silt.

Columns have also been "righted" by means of steam-hoists pulling upon the side to which the cylinder inclines; by screw-jacks, hydraulic-rams, and other powerful lifting tackle ; by additional weighting upon the higher side ; excavating the ground to a slope on the inclined side to the lowest water level, if any part of the ground upon which the pier rests is dry at any time, and by propping up the leaning side with rails, sleepers, or other hard material so as to cause a large firm wedge-shaped mass to press against and support it. Hollow pointed perforated iron tubes may be driven in, or sunk by the water-jet system, around and close to the cylinder on its higher side, to loosen sand and such soils and to diminish surface friction by the aid of water discharged through them. They can be withdrawn and inserted as desired, and may be useful adjuncts not only in "righting" but also in sinking a cylinder.

The internal raising of excavation in one direction, and the vibration so caused, have been found to make a cylinder incline towards the source of power of the hoisting apparatus of dredging machinery. This can be soon noticed, and counteracted either by changing the position of the machinery frequently, by propping, or by other means.

CHAPTER VIII.

KENTLEDGE.

THERE are several different ways of arranging the kentledge or the weights for sinking a cylinder. Its cost often amounts to a considerable sum, and much time is required to remove it. Perhaps the cheapest temporary kentledge is obtained by putting across the top of the cylinder some balks, and evenly and equally packing upon them medium-sized stones, afterwards used in the masonry of the bridge, or by placing rails or pig iron upon timbers resting upon the horizontal flanges or top, leaving sufficient space for working and excavating operations. The weights should be so arranged that they do not interfere with the internal excavation; they should be equally distributed throughout the cylinder to ensure uniform sinking; and care should be taken that the weight on the iron rings and flanges is not excessive, and that the temporary loading does not cause a cylinder to tilt.

In placing kentledge upon the top of a cylinder the centre of gravity is raised, and the column may not be stable. It is sometimes laid upon stages slung within the cylinder, as being not so likely to tilt the column as when placed outside, and as being more easily thrown off when the requisite depth is reached. A method that has been adopted is to have weights cast to the form of the cylinder, and so made that they fit into the concave side of the column and rest on the horizontal joint-flanges. The weights are cast in convenient sizes, such as 6 or 7 ft. in length, 1 ft. in height, and 6 in. in thickness. The dimensions most convenient depend upon the size of the cylinders and the width of the flanges. The weights can be lowered down the interior of the cylinder by a crane on the staging. All the lengths of the column by this system are weighted with the exception of the cutting-ring and the next above it, and if the weights are properly placed they mutually support each other as they act as arch stones. From 10 to 30 tons of kentledge can thus be readily placed out of the way of operations on a 6 or 9 ft. length of cylinder ring of ordinary diameter. These weights can remain upon the cylinder during sinking, being removed as the hearting reaches them, and there is no danger of tilting by unequal distribution of weight which occurs when the load is on the top. It seems to be considered, if the cylinders to be sunk are of considerable height and many columns have to be erected, that the advantages gained by having the weights cast amply compensate for the extra expense of such special kentledge, the latter being made to fit all the cylinders and being sold after use. Sir Bradford Leslie, M.Council Inst.C.E., introduced a system of weighting cylinders by means of a water-tank, which could be filled by pumps in from ten to fifteen minutes, and when empty was lifted on the top of the cylinder. When a column had sunk sufficiently for another segment to be attached, the tank was emptied, swung off, let down on the top of the next length, and refilled. The same tank does for any number of cylinders, the delay and expense in placing rails and weights are obviated, and the load is equally distributed all round the cylinder ring. It was found that loading a cylinder with rails and iron took twenty hours; fixing and filling the tank occupied but one hour. This system is decidedly preferable, if tackle is at hand for swinging the tank off and on, to weighting by rails, etc.; but it has not the advantage that inside kentledge has, namely, that of distributing the weight over the height of the cylinder and practically maintaining its centre of gravity unaltered.

The comparatively new method of inside kentledge will now be examined. As rapidity is generally to be desired in sinking cylinders, and is often imperatively necessary, the value of using a casing of the permanent hearting as a load for aiding sinking operations can hardly be over estimated, for it can be done with but little extra expense to the ordinary cost of the hearting. It leaves the top of the cylinder entirely free to receive an additional load; it causes the centre of gravity of the cylinder to remain low, and saves the trouble, waste of time, and expense of constant stacking and re-stacking weights; and part of the casing or hearting is built on land and is not subject to any pressure of water before it is set. The chief precautions to observe in using as kentledge such an internal casing are not to make it too heavy for regular sinking, or to cause a cylinder to run down so quickly as to crush the cutting edge should it come in contact with boulders, or a hard stratum, or rock; to be sure that the casing fits the flanges of the cylinder, whether it is made of masonry, brickwork, or Portland cement concrete, the latter material being preferable, as it can be made to fill spaces between bolts, ribs, and other projections, but time should be allowed for it to set; and sufficient open space must always be left for excavating and the other ordinary operations of sinking.

Experience in cylinder sinking indicates that it is prudent to, as it were, pull a cylinder down, as well as to weight it, which can be done either, (1) by building up a portion of the hearting on an annular ring, having a sufficient opening for purposes of excavation; (2) by specially cast kentledge placed on the horizontal flanges; (3) by having a stage loaded with weights, and slung from the flanges at intervals. The load is then much more equally distributed than can be the case in any system of top loading, the compressive strain on the rings is reduced, which lessens the possibility of fracture of the rings, and also conduces to prevent tilting of the cylinder, for the effect of weighting at intervals from the top of the cutting ring is to assist vertical sinking, provided the weights are equally distributed. In small cylinders internal loading may be difficult to arrange, because of the space being required for purposes of excavation, but in cylinders from about 9 to 10 ft. in diameter, a casing of the permanent hearting of the cylinder can be built so as to leave sufficient space for excavating or dredging machinery, and general operations. Much time and considerable expense is saved by using a casing of the permanent hearting for kentledge, instead of temporary loading, for it obviates the removing and restacking rails, stones, or other weights, each time a ring of the cylinder has to be added, which is always a slow, expensive, and weary process.

As a rule, with but few exceptions, large cylinders are to be preferred to small columns. The reasons have been given. An objection sometimes raised against large cylinders is that more kentledge is required to sink them, which undoubtedly is true if considered merely from the view of the gross weight requisite to sink a cylinder, but is not if considered from that of the frictional surface resistance that has to be overcome as compared with the area of the base in order to cause a column to sink; see Chapter IV., on Surface Friction, pages 33 to 39 inclusive, in which it is named that "in a cylinder of large diameter the proportion of its circumference to the area of the base is small. On the contrary, where a cylinder is of small diameter, the circumference is nearly equal to the area of the base."

For instance—Supposing it is found that a 12 ft. in diameter, 1½ in. in thickness cylinder is required under each main beam of a bridge, or two for a single line railway-bridge.

The area (see Table A, column 2, Chap. II., page 18) of two 12 ft. cylinders is $113 \cdot 10 \times 2 = 226 \cdot 20$ sq. ft.

The surface area of two 12 ft. cylinders, 1½ in. in thickness (see Table B, column 4, Chap. II., page 20) is $38 \cdot 48 \times 2 = 76 \cdot 96$ sq. ft. per lineal foot of the height of the cylinder.

Now an area of 226·20 sq. ft.=two 12 ft. cylinders, must be provided in order to safely support the weight of the cylinders, superstructure, and rolling load.

Assume that 12 ft. in diameter cylinders are considered to be too large, and that 8 ft. 6 in. in diameter columns, 1½ in. in thickness, would be handier to sink, are thought to give a better base, to reduce the gross resistance of the surface friction of each cylinder, and therefore to require less kentledge than a larger cylinder, how many would be required?

$$\text{Square feet.}$$
$$\text{The area of two 12 ft. cylinders} = 226 \cdot 20$$
$$\text{The area of an 8 ft. 6 in. cylinder} = 56 \cdot 75$$

consequently $\dfrac{226 \cdot 20}{56 \cdot 75}$=say, four 8 ft. 6 in. cylinders would be required.

The surface area of four 8 ft. 6 in. cylinders, 1½ in. in thickness, per foot of the height is (see Table B, column 4, Chap. II., page 20) 27·49

×4=109·96 sq. ft.   The surface area of the four 8 ft. 6 in. in internal diameter cylinders per foot of their height would therefore be $\frac{109·96}{76·90}$
=1·43 times greater than the surface area of the two 12 ft. cylinders, and the *total* weighting would be in round numbers 1¼ times more than that of the two 12 ft. in diameter cylinders, although one 8 ft. 6 in. cylinder would only require $\frac{27·49}{38·48}$=0·71 of the kentledge requisite for sinking a 12 ft. cylinder under similar circumstances.

Assume the 12 ft. in diameter cylinders have to be sunk to a total depth of 40 ft. in soil having a frictional resistance of 280 lbs. per square foot, or ⅛th of a ton. How much kentledge would be required when the cylinder had reached 30 ft. in depth?

The frictional resistance of the 12 ft. in diameter, 1½ in. in thickness cylinder,

Surface Area.   Depth.   Square Feet.   Ton.   Tons.
= 38·48   ×   30   =   1,154·40   ×   ⅛   =   144·30

The area of the cutting edge of the 12 ft. cylinder, 1½ in. in thickness which is here taken to be flat, as being sufficiently near for purposes of calculation, although of course it would taper, is 38·09 sq. ft.×1½ in.= 4·76 sq. ft. This area has to be thrust through the earth, unless the excavation *always* extends under it, which is unlikely. What weight would be required to make it sink? Although it might be estimated on the surface, at such a depth as 30 ft. it cannot be known, but it may be approximately determined by deductive reasoning. With a cutting ring tapered to about ½ in. in thickness, and the soil bared as the cylinder descends, it would probably not exceed the normal pressure of the soil, which would be at a depth of 30 ft., assuming it was earth weighing 0·055 ton per cubic foot; 30×0·055=1·65 tons. Consequently the force required under the conditions named would probably be, area, 4·76 sq. ft.×1·65 tons=say, 8 tons.

The total resistance to be overcome by the kentledge, disregarding fractions, would therefore be:

		Tons.
Frictional resistance = 144 tons + cutting edge resistance = 8 tons = … … …		152
DEDUCT.—The weight of the cast iron rings of the cylinder 12 ft. in diameter, 1½ in. in thickness (see Table B, column No. 3) …	Ton. 0·95	
Add 20 per cent. for flanges, ribs, etc. … …	0·19	
	1·14	
1·14 × 30 =		34
Weight of kentledge required =		118

REQUIRED.—The thickness of an annular casing of Portland cement concrete to produce this weight.

A cubic foot of concrete weighs, say, 136 lbs. = say, $0\cdot 06$ ton $\frac{118}{0\cdot 06}$ = the total cubic feet of concrete required = 1,966 cub. ft., or per foot of height $\frac{1,966}{30}$ = say, 66 cub. ft.

The internal area of a 12 ft. cylinder = 113·10 sq. ft.

DEDUCT.—Portland cement concrete area required per foot of the height of the cylinder = 66·00 ,,

Leaving 47·10 area in square feet or open space left for excavating operations.

By looking down column No. 2, Table A, the areas near this are those for a 7 ft. 6 in. and 8 ft. cylinder. Take the former, which will allow sufficient space for excavating operations. Consequently the thickness of the annular ring would require to be

$$\frac{(12'\,0'' - 7'\,6'')}{2} = \frac{4'\,6''}{2} = \text{say, } 2'\,3'',$$

to produce the necessary deadweight.

If four 8 ft. 6 in. in diameter cylinders were adopted instead of two 12 ft., the permanent hearting could hardly be used as temporary kentledge, for there would not be a convenient open space left for excavating and hoisting apparatus, and general sinking operations, as the following calculations show :—

The surface area of an 8 ft. 6 in. cylinder, 1½ in. in thickness, per foot of the height (Table B, column 4), is = 27·49 sq. ft.

                        Surface area. Depth. Sq. ft. Ton. Tons.

The frictional resistance = 27·49 × 30 = 824·7 × ⅛ = 103·09
Cutting edge resistance to penetration, say = 5·57

                                                        108·66

DEDUCT.—The weight of the cast iron 1½ in. rings of the 8 ft. 6 in. in diameter   Ton.
cylinders ... ... = 0·68
Add 20 per cent. for flanges, ribs, etc. ... = 0·14

                                                        0·82

0·82 × 30 =, say, 24·66

Weight of kentledge required   84·00

Proceeding as for the 12 ft. cylinders the total cubic feet of Portland cement concrete required ... ... $= .\tfrac{84}{85} = 1,400$ cub. ft
or per foot of the height of the cylinder $= \tfrac{1400}{30} = 47$,,
The internal area of an 8 ft. 6 in. in diameter cylinder $= 56\cdot 75$ sq. ft.

DEDUCT.—Portland cement concrete casing area required per foot of the height of the cylinder $= 47\cdot 00$,,

9·75 area in square feet left for excavating and sinking operations.

A 3 ft. 6 in. in diameter opening would approximately give this area, therefore, the thickness of the annular ring would be,

$$\frac{(8'\,6'' - 3'\,6'')}{2} = \frac{5'}{2} = 2'\,6'',$$

to produce the necessary deadweight.

The ratio of the diameter of the 12 feet cylinder to that of open cylindrical space left ... $= \dfrac{12}{7\cdot 5} = 1\cdot 60$

Ditto, ditto, of the 8 ft. 6 in. cylinder ... $= \dfrac{8\cdot 5}{3\cdot 5} = 2\cdot 43$

an important difference, for according as it becomes greater so will the difficulty of excavating over the area of the base of the cylinder increase, and in any but very movable soil in the 8 ft. 6 in. diameter cylinder it would be most difficult to excavate in tenacious earth, or to make the earth fall into a central hole made below the base of the cylinder. That is one objection. Another is that of the ratio of the open area to the depth in each case. Taking the depth of 30 feet:

In the 12 feet cylinder, the open cylindrical space is 7 ft. 6 in., or $\dfrac{7\cdot 5}{30} = \tfrac{1}{4}$ of the height.

In the 8 ft. 6 in. cylinder, the open cylindrical space is 3 ft. 6 in., or $\dfrac{3\cdot 5}{30}$ say, $\tfrac{1}{9}$ of the height.

The dimensions of the interior unobstructed area will be chiefly regulated by the size of the dredger machinery and the depth of the foundations. For tenacious earth, such as the clays, in which it is desirable to have a heavier dredger in order to help penetration than in light, loose soil having little tenacity, open space is of importance, and the larger it is the better, and it should increase according to the depth as the normal weight of the soil is likely to make it more compact, and therefore the distance of the extended edges of the scoops of the dredger from the metal ring of the cylinder should be less as the depth increases, for the ground will generally be harder, and control of the apparatus not so easy as at less depths. Perhaps a minimum free working area of 5 feet diameter for excavating operations is to be preferred in loose soils, and 6 ft. to 6 ft. 6 in. in tenacious soils. Taking these as *minimum* dimensions, sufficient deadweight for sinking from a casing of the permanent hearting, without other additional means, can, therefore, not be obtained in cylinders of a less diameter than about 9 to 10 ft. respectively, when they are sunk to ordinary depths. If the uncovered space commenced with a minimum of 5 ft. diameter for a depth to be sunk of 25 ft., it is desirable to increase it as the square root of the depth to be sunk ; thus for 25 ft. in depth it would be 5 ft. ; for 50 ft., say, 7 ft. ; for 100 ft., 10 ft. In all cases the open area should be as large as possible, sufficient thickness being allowed for the annular ring of permanent hearting to act as kentledge.

When a free working area is provided for easy lowering and hoisting, at little depths up to about 30 ft., the excavating apparatus may be readily controlled, but with increasing depth it becomes important to have more room in which to control it, and to provide for a diver to descend in case it becomes fixed or broken, or will not penetrate, especially at the base of the cylinder, and for this reason every care should be taken that no projections are left in the rings for the excavating machinery to be caught, consequently the annular ring supporting the casing of the hearting should have a triangular end with the point downwards, so that the dredgers, if they come in contact with the sides of the casing, will slide up on being hoisted. In case of the scoops holding a boulder or being much open as in catching a log, it is necessary that the unobstructed space is sufficiently large for them to be hauled up when they are fully extended from any cause, and especially so in very deep foundations. If the size of the cylinder will not admit of a sufficiently thick casing to produce the necessary deadweight for sinking it, as wide a casing as convenient can be adopted, but it is not advisable that it should be of less thickness than 1 ft. 3 in. to 1 ft. 6 in. for depths not exceeding 25 ft., and 1 ft. 6 in. to 2 ft. are preferable *minimum* thicknesses for foundations sunk to any ordinary depths. However, the deadweight thus gained will much reduce the quantity of movable kentledge required and steady the cylinder during sinking operations.

The necessary quantity of kentledge will vary with the nature of the soil through which the cylinder has to be sunk, and according to its size and depth in the ground. For cylinders of from 8 to 14 ft. in diameter, and sunk to ordinary depths up to about 60 ft., from 50 to 400 tons may be required. The deeper a column is sunk, the more it will want, the weight increasing as the surface area in contact with the earth becomes greater. In mud, 50 to 100 tons will probably be sufficient for cylinders of moderate diameters. In sand and tenacious soil, from 75 to 300 tons, and in tenacious and adhesive earth, 100 to 400 tons in addition to the weight of the cylinder, and the plant placed thereon; but the load required is much influenced by the continued rapidity and regularity of the excavating operations. When the compressed air system is used, in calculating the quantity of kentledge, the lifting power of the air-pressure must be added to the weight required as it must be counteracted. Reference to the notes on "Surface Friction" (Chapter IV., pages 33 to 39) will enable an approximate estimate to be formed. Some methods of lessening the surface friction, and procuring easy sinking have been previously mentioned.

CHAPTER IX.

Hearting.

WITH regard to the hearting or internal filling of a cylinder bridge-pier, it may be made to fulfil a three-fold purpose, namely :—
1. To support the weight of the superstructure and rolling load, and that of the pier, *i.e.*, the iron casing and its own weight.
2. To water-seal the bottom of a cylinder.
3. To act as a weight to sink a cylinder.

It is not within the scope of this book to refer to the fabrication or construction of the material of which the hearting may be composed. It usually consists of Portland cement, concrete, brickwork, or masonry, sand being occasionally used for filling up cavities at the base, and also as hearting in wells forming quay walls when the insistent weight is moderate. The first question to determine is, which is the best and cheapest hearting? It cannot be said that any material is the best to use under all circumstances, for rapidity of execution, cheapness, durability, homogeneity, and strength, have all to be considered.

Bearing in mind that the hearting should be a thoroughly homogeneous mass, Portland cement concrete is to be preferred on this account to brickwork or masonry, for the mortar holding them together has generally a cementitious strength below that of the bricks or stones; therefore a strong Portland cement mortar having a high cementitious value should be employed, its maximum strength being reached and maintained within a reasonably short time of its being mixed, but care should be exercised that the Portland cement has not too much lime in it in order to produce an early high tensile strength, for, if the concrete is

F

immersed, the excess of lime will sooner or later become slaked by moisture, and then the mass will expand or crack, and the concrete probably be more or less disintegrated. Lime mortar should not be used in the hearting, as the decrease of strength, durability, and cementitious value is considerable.

It is important that the deposition of the concrete hearting be done equally; it is therefore desirable that it be put in the cylinder in layers not exceeding 1 ft. in thickness. Objections have been raised against the adoption of concrete hearting when it has to be cast in among an entanglement of temporary timbers, as then cavities are more likely to occur than when brickwork or masonry in cement are used, which has to be carefully built; however, if ordinary care is taken in depositing the concrete, it may be said that a few small cavities are certainly not worse than a mass consisting of a conglomeration of hard materials possessing great strength and durability, but joined together by a weaker and less durable substance.

At the Dufferin Bridge, over the Ganges, at Benares, it was found that the weight of the pier superstructure caused a settlement of nearly 3 in., and this was chiefly attributed to cavities occurring in the hearting. The plan was then adopted of filling in the lower part of the cylinder or well to the top of the conical base plate with sand before depositing the concrete hearting. The sand settled more closely under the slopes at the base than when concrete was used, and much reduced the settlement, which was by no means excessive.

As foundations have been repaired by means of Portland cement grouting, consisting of equal parts of Portland cement and clean sharp sand, forced by a hand pump through pipes having a diameter of from $1\frac{1}{2}$ to 2 in. in a continuous stream of 1 or 2 cubic yards for each pipe, time being allowed for the grout to set before resuming operations, the grout being injected until no more can be received, it might be used for filling cavities round the cutting edge of a cylinder. The writer's book "Notes on Concrete and Works in Concrete" contains brief practical information as to the composition, air-slaking, testing, proportions of the ingredients, mixing, and deposition, etc., of concrete.

The earth at the base of a cylinder must be cleared and levelled by divers or otherwise before the hearting is deposited, and the concrete should be evenly spread and trodden down by a diver or other means, and until the hearting has thoroughly set it should be kept entirely free from dripping water, and no pumping should be allowed when concrete is being lowered through water. As the concrete in a cylinder is in a considerable mass, and is often covered up as soon as deposited, it is imperative that only quick-setting material be used, and that it possesses increasing powers of resistance.

At the New Tay Bridge, in order to provide against the possibility of the wrought iron cylinder rings of ⅜ in. in thickness, and 16 ft. 6 in. in diameter, enlarged to 23 ft., perishing from corrosion, the Portland cement concrete hearting is encased in a ring of hard brickwork set in Portland cement as a second shield of protection.

It has been found that concrete will not set under hydrostatic pressure, and that the water pushes its way through the interstices of the stone before the cement has time to harden sufficiently to resist it. And with compressed air it generally happens that the upper surface in contact with air-pressure sets quickly, so that the rest of the mass derives very little or no benefit from the air-pressure, unless means are taken to bring it in contact, or opposition to, the force of the water. Small vertical pipes leading downwards to the bottom of the concrete, and placed within 1 ft. of each other over the whole area of the column, have been used to obviate this difficulty. It is advisable in all cases to test the setting powers of any concrete to be used under a similar pressure to that it will have to sustain when being deposited. General experience seems to show that concrete laid down under compressed-air sets quicker and slightly increases in strength, provided it is deposited in thin layers which allow any excess of water to escape.

In cylinder piers no danger is to be apprehended from the unequal contraction of the iron rings and the concrete, if ordinary precautions are taken. There are numerous iron cylinder bridge piers in all parts of the globe filled with concrete, subject to temperatures ranging from 150° F. to a few degrees below zero, which have stood perfectly under all conditions of traffic ; but it should not be forgotten that the unequal contraction of cast or wrought iron and concrete by cold, and the freezing of water in a cylinder, may produce internal strains on the rings, for the occasional bursting of water-pipes shows that the limit of elasticity of cast iron is sometimes insufficient to allow the metal to make the necessary expansion or contraction, hence the exclusion of all water inside the cylinder is of importance, and consequently the joints of the cylinders should be caulked, or rendered water-tight in some manner, so as to prevent any strain from the expansion of water on freezing, and also aid any pumping out of water before the cylinder is filled with the hearting. At the South Street Bridge, Philadelphia, U.S.A., where, as usual, no allowance was made for the contraction of the cylinder during frost, four or five rings split horizontally or vertically during the first winter, the worst crack, a vertical one, opening nearly ¼ in. A lining of wooden staves is recommended to prevent this occurrence, as the wood would be compressed sufficiently to relieve the strain on the metal. It is advisable to make the top of cylinders watertight so as to prevent any percolation of water between the hearting and the rings, which may cause them to crack in frosty weather, and any holes made in the rings for the purpose of aiding sinking operations should be carefully filled. Anchor bolts with washers are occasionally embedded in the concrete hearting, and pass vertically through the whole of it from the base to the capping-piece on which the girder rests, with the view to produce additional strength and solidity.

With respect to water-sealing a cylinder by depositing some of the lower portion of the hearting, unless it is composed of Portland cement concrete it cannot be done, but by the special adaptability of that material for air-sealing a cylinder, the compressed-air system will frequently not be required. In order to water-seal a cylinder, two chief points are presented for consideration, the proportion of the aggregates to the cement and the required thickness of the seal. A very small head of water will cause it to rise through concrete made of 10 of aggregates to 1 of Portland cement, but a considerable head is required to make it percolate through such a mixture as 6 to 1 when it is properly proportioned and mixed with a view to solidity and imperviousness. With this object the Portland cement concrete seal might be made of 2 of sand to 1 of Portland cement for the first few feet, and then a 4 to 1 to 6 to 1 mixture, or 4 to 1 for the first few feet, and upon it a 6 to 1 mass, thus making a nearly impervious surface in contact with the water, and causing the concrete to be in a condition to readily spread and fill any cavities at the base, and be well distributed under the cutting edge.

The strength and thickness of the seal must be increased according to the head of water and porosity of the soil, but no rules can be fixed, for the conditions vary. From 5 ft. to 15 ft. of Portland cement concrete will generally seal a cylinder. The concrete should be allowed a week or so to set thoroughly before the water is baled or pumped out, and should be properly trimmed and trodden, or gently beaten solid, and if it has to be lowered through water it should be made richer than if it had to be deposited in the air, to compensate for any cement that may be washed out during lowering it.

When a depth of about 15 ft. of concrete has been deposited in a cylinder through deep water, the following plan is sometimes adopted: A disc of planking from 3 to 4 in. in thickness, and a few inches smaller in diameter than the cylinder, is let down upon the surface of the concrete and weighted; the space between the edge of the disc and the sides of the column are next filled in with wooden wedges driven in by divers. The concrete is thus prevented from being disturbed by the pressure of water underneath, and the water can be baled out without causing a flow or material agitation. This method is especially useful in bridge-well sinking, and where the head of water is great. If the concrete is not weighted or prevented from moving, it may be blown up by the pressure of the water underneath. Weighted thick tarpaulins or any practically impervious and suitable material can be used for small depths. The Portland cement concrete can be put in by divers, or by the usual shoot-boxes in two pieces, hinged and fastened by a catch which can be released on pulling a rope attached to it, or in bags, but the hearting must be made to act as a monolith. Where the cylinder is erected on dry land, the concrete can be raised and lowered by an endless ladder or other usual hoisting and lowering apparatus, great care being always taken that the concrete is not thrown down from a height, but gently emptied upon the base, or much of the cement will be washed out ; the heavier material will fall quickest, and the concrete be unequable in strength and character.

A few thicknesses of the Portland cement concrete water-seal that have been sufficient to stop the ingression of water, are named to illustrate the variableness of the required mass. In the case of a cylinder 14 ft. in internal diameter, the greatest depth of water being 13 ft., the depth sunk through the river bed 38 ft., and the ground gravel and silt, 4 to 5 ft. of 4 to 1 Portland cement concrete, when set, sufficiently sealed a cylinder so that it could be pumped dry. In another example, a 12 ft. in diameter cylinder required 8 ft. of similar concrete to seal it sufficiently for the water to be ejected. In each case seven days were allowed for the concrete to set. In another instance, where a Portland cement concrete composed of 4 of stone, $2\frac{1}{2}$ of sand, and 1 of Portland cement was used, no less than 18 ft. of concrete had to be deposited, the depth of water being 50 ft. The necessary thickness varied very much from a maximum of 18 ft. to a minimum of 4 ft. The soil was sand for a depth of about 35 ft., and clay and sand and clay for 20 ft.

Some of the causes influencing variations in the required thickness of the seal are, apart from the depth or head of water, the relative porosity of the earth that is penetrated ; the close contact of the outside soil with the surface of the iron rings ; fissures or depressions in the bed of the river ; considerable range of tide, which tends to keep loose soil in a state of unrest and insolidity ; the inclination of the strata which may localise and augment the pressure ; the area of the cylinder, the required thickness being greater as it increases ; and the composition and character of the seal.

CHAPTER X.

THE COMPRESSED-AIR METHOD OF SINKING CYLINDER

DR. POTT'S vacuum principle of sinking cylinders, which was practically introduced in 1839, has been generally abandoned in favour of the compressed-air method. This latter system was first adopted at the Rochester Bridge, England, in 1851-2 ; but air-compressors were first practically used by Smeaton in 1788, etc., at Ramsgate Harbour, England. They were there employed for diving apparatus.

The plenum or compressed-air method of sinking is all but certain in its action, which can hardly be said of any other system if obstructions are likely to be met with, or in a difficult situation ; it can be further aided by the use of mechanical pressure or weight ; however, provided it has been ascertained that no obstructions will be encountered, some of the other methods of sinking previously described may be used with economy.

In the early examples of the use of compressed air in sinking cylinders the whole of the cylinder from the air-lock downwards was filled with compressed air ; but the system of a working chamber at the bottom, with a communication pipe sufficiently large for men and materials to pass, and an air-lock at top, is now generally used, the space unoccupied by the compressed air being filled with the atmosphere, water, masonry, brickwork, or concrete, which assist in sinking the cylinder. Many improvements and modifications have been from time to time introduced, such as placing the air-lock immediately over the working chamber at the bottom instead of at the top of the cylinder, and by the working chamber, air-lock, and column being suspended by links, and raised as the pier is built, thus requiring no iron skin to the pier. The latter method is practically a funnel-shaped diving bell. By the use of compressed air such heavy weights as that of an iron cylinder, which will frequently amount to from 40 to 100 tons, may be manipulated during descent with ease, by the aid of simple and inexpensive apparatus and a few men, who should be experienced, none but skilled foremen being employed.

The main points in sinking by means of compressed air are to supply sufficient air for the expulsion of the water, and for the men in the working chamber ; to provide for the ready entrance and exit of the men, for the introduction of plant and the hearting, and for the removal and discharge of the material excavated. In sinking cylinders by the compressed-air method, if the depth is considerable, the pressure of the air necessary to exclude the water may be sufficient to overcome the weight of the cylinder and the surface-friction ; if so, and provided the cylinder is not weighted, it will be lifted until the pressure of the air and the weight of the cylinder and the surface friction balance ; and care must be taken that the pressure does not blow up the top of a cylinder or floor of a caisson.

In order that a cylinder may not sink without some air being let off, or the cutting edge being undermined, its total weight when loaded must be less than the surface friction in addition to the flotation power and the resistance of the ground to penetration by the cutting ring. For motion to take place, the effective air pressure in addition to the surface friction and the resistance of the ground to penetration by the cutting ring, must be slightly less than the weight of the loaded cylinder. The reading of the gauges should be recorded when motion commences, and when it is arrested. By gently opening the safety-valve, or by pumping in more air, the pressure may be lowered or raised as desired. On the air-pressure being removed, or greatly lessened, a cylinder will in ordinary soil go down many feet; a sudden sinking as much as 20 ft. has been caused by a large reduction of the air-pressure, the surface friction being quickly overcome. When such a precipitous depression takes place it will generally be found that the earth in the interior of the cylinder will rise very considerably more than the depth to which the column has sunk; in many instances the excess of the soil in the column has been found to be as much as from 50 to 100 per cent., in very loose soil it would probably be greater, and may, perhaps, entirely fill the cylinder. Such sudden sinking is not economical, or to be desired.

It is doubtful whether a reliable comparison of the cost of the compressed-air system with other methods of sinking can be made, as all the conditions and circumstances require to be exactly alike, and as this only occasionally occurs, the difficulty is to make correct additions and reductions. However, with great ingenuity, attempts have been made, with various results, to establish the depth at which each system is the most economical and to be preferred. The least depth for the economic adoption of compressed air has been stated to be as little as 16 ft. of water, and again as 25 ft. The determination is replete with difficulties, for it is not to be expected that the exact cost in detail of every item of expenditure will often be stated, and without it any conclusions that may be drawn will be misleading. Doubtless a reliable comparison can be occasionally made, but the point is, is it of general application, and would anybody be willing to be bound as an engineer, bridge-builder, or contractor to erect a bridge-pier according to the estimates so formed. Most probably not. To be financially interested in the correctness of an estimate is a somewhat different operation to that of having but a purely scientific regard for it. Comparative estimates made from fifteen to twenty years ago between the cost of sinking by compressed air and by other means, are hardly applicable at the present time, for the improvements made in dredger-excavators during about that period have been most marked, whereas the general apparatus necessary to be used in adopting the compressed-air system has been but slightly improved, and been more confined to small details than improvements made with the view of lessening the cost of that method of sinking cylinders or caissons. A few of the advantages of the compressed-air system may be said to be :—

1. The excavation can be done on dry land, and therefore the much greater certainty of sinking a cylinder.
2. The easy examination of boulders, logs, or other obstructions encountered in sinking.
3. The possibility of using means and methods of excavating the core, and especially the removal of obstructions in sinking, not economically available except on dry land.
4. Greater control of the cylinder from the power to increase or lower the air pressures, and therefore the additional means afforded of causing a hanging cylinder to sink.
5. The laying bare the base of the cylinder for the hearting, which need not be deposited through water.

Some of the disadvantages may be considered to be :—

1. The deleterious effect of the compressed air on the men when the pressure is more than $1\frac{1}{2}$ to 2 atmospheres above the ordinary atmospheric pressure, and the consequent shorter hours of labour, which must be decreased with the increase of pressure.
2. The liability of sudden and fatal accidents occurring.
3. The prudential necessity of having much of the compressed-air machinery in duplicate.
4. The employment of additional skilled labour.
5. The increased plant required, such as air-compressors, pumps, air-locks, working-chamber, light, boilers, engines, smithies, etc.
6. The expense of making the cylinder as air-tight as possible.

The question of relative speed of excavation is not considered, as the size of a cylinder and the working area will influence that operation.

CHAPTER XI.

LIMITING DEPTH, AIR-SUPPLY, AND LEAKAGE.

IN using the compressed-air system for depths such as 80 or 100 ft. there is difficulty in making the cylinders air-tight without special care in construction, and at the depth of 100 ft. below the water level it is dangerous to the men unless special precautions are taken. Many examples of cylinders sunk 80 to 85 ft. below the water-level by the compressed-air system exist, and at the St. Louis Bridge the men descended to a depth of 120 ft., which may be considered as about the maximum advisable or even practicable depth; the maximum air-pressure above the ordinary atmospheric pressure was nearly 52 lbs., or about $3\frac{1}{2}$ atmospheres. It is certainly very questionable whether the compressed-air method is the best system that can be used for depths greater than from 80 to 100 ft. below water-level, nor is it usually economical for depths less than about 25 ft. below water, except under special circumstances, but the depth for its economic adoption is not easy to determine generally, as has been mentioned.

There is considerable waste of air in the pneumatic process, through the air escaping under the bottom of the cylinder, and by leakage, and much more power is required than that shown by actual work. Careful construction of the cylinder, particularly of the joints, and caulking the latter with tarred oakum, or other approved water-tight preparation, will save expense by preventing leakage of air and the ingress of water. Sometimes it has been found necessary to line the interior with cement to make the cylinder sufficiently air-tight; but this is seldom requisite if due care is taken in the construction of the cylinder.

Authorities somewhat differ as to the actual quantity of air consumed by a man, but 220 to 240 cub. ft. per hour is sufficiently near for practical purposes, and about one-twentieth of this amount, or 11 to 12 cubic ft. per hour for an ordinary candle. This is the net quantity of air required without allowing for leakage. There is the leakage from escape of air under a cylinder, through the column, air-pipes, lock, etc., and the constant loss of air during the passage of workmen and material through the air-lock. The loss from leakage will almost always determine the necessary supply, hence the importance of making an approximate estimate of its amount. Unless the power of the air apparatus greatly exceeds the amount of air theoretically required, it will be necessary to continuously keep the pumps at work; under ordinary circumstances and without great care, it is not advisable to stop them. The leakage through and under the column will be found to be much greater than that of the air-pipes and lock, and the point of leakage can generally be discovered, and in great measure stopped, by attention to the joints, and by covering them with some impermeable material, such as tempered clay, which will be forced into any cracks or crevices by the pressure of the air, if considerable. Leakage may also be expected where the roof of the cylinder joins the rings. So long as the water is above the level of the bottom of a cylinder, the air will be prevented from escaping at the base; but if the sinking of the column is arrested during the progress of the excavation, the water may be entirely driven out of the column; there will then ·be an escape of air under the cutting edge, the effect of which is to make the soil around the cylinder loose and spongy, thereby lessening the surface-friction, but increasing the liability of the column to tilt; still, in clay soils, notwithstanding the earth being excavated under the cutting-ring of a cylinder, it may not sink, and in addition to extra loading it may be necessary to raise the air-pressure half or three-quarters of an atmosphere above that required to exclude the water so as to disturb the soil pressing upon the cylinder rings, to reduce the frictional resistance, and to cause the cylinder to descend.

The loss of air during the expulsion of the water from a cylinder is usually considerably less than when it has been expelled, and excavating operations are in progress; and it will vary according to the workmanship, the nature of the soil, the pressure, and whether the sinking of the column is carefully arranged so that there is not much loss of air under the cutting ring. In coarse gravel and clay the loss may be said to be, all other conditions being similar, from 8 to 9 per cent. less than in fine gravel and sand. Where there is leakage of air through any timberwork, if not through the wood, it can generally be lessened by watering, which will cause the wood to swell.

Both the air-lock and the shaft of an iron cylinder can be made almost air-tight; but it is nevertheless necessary that air be constantly discharged into the cylinder so as to keep it comparatively fresh and pure to live in, or respiration cannot be efficiently performed. The air pressure should not be excessive, or the men may be inconvenienced, but it should be sufficient to keep the air in a fresh and pure state, and so that there is no want of air. This fresh air is wasted regarding it from the point of view of pressure, but such extra air has been utilised by an arrangement by which it is allowed to escape through pipes, and to carry sand and loose soil with it to the top of the cylinder.

It has been found that when the natural skin, or surface left in casting has been removed from cast-iron, water under pressure of about 3 to $3\frac{1}{4}$ tons per square inch will pass through the pores of the iron; such pressures, however, can never occur in cylinder sinking. As air under a pressure of about 40 lbs. per square inch, or 2·72 atmospheres, will penetrate most wood, if a timber roof on a cylinder should have to be used in the pneumatic system, it is necessary to coat the wood with some air-tight preparation, such as resin or euphorbia-juice. The experiments of Professor Doremus show that air can be easily forced through sandstone, brickwork, and unglazed tiles, etc. A depth of a few feet of water produces sufficient pressure to enable it to percolate red sandstone. In using the compressed-air system with brickwork and masonry, it is necessary to coat them with an impervious preparation. or neat cement.

To calculate the pressure of air required to balance the water-pressure the following formulæ may be found useful:—

Let D = the depth in feet of the foundation below water level.

„ A = the ordinary atmospheric pressure=say, 14·71 lbs. per square inch.

> NOTE.—Although the pressure of the atmosphere varies between, say, 13·65 lbs. per square inch at sea level, and 15·06 lbs., 14·71 lbs. is generally taken, which equals 29·92 inches of mercury.

„ N = number of atmospheres above the ordinary atmospheric pressure, or gauge pressure.

„ W = weight of a column of fresh water 1 in. square and 1 ft. in height=0·433 lb.

„ P = pressure of air in lbs. per square inch required in the cylinder to balance the water pressure, in addition to the ordinary pressure of the atmosphere.

„ Pp = Pressure required above a vacuum in lbs. per square inch, then,

$$P = WD. \quad N = \frac{WD}{A}, \text{ or as } WD = P, \quad N = \frac{P}{A}.$$

$$Pp = P + A.$$

Example.—Let the greatest depth below the surface of the water= 50 ft. Required P, Pp, and N.

$$P = WD = 0·4333 \times 50 = 21·66 \text{ lbs.}$$

$$Pp = P + A = 21·66 + 14·71 = 36·37 \text{ lbs.}$$

$$N = \frac{21·66}{14·71} = 1·47.$$

The total pressure per square inch above a vacuum in the cylinder would therefore require to be ... ... Gauge Pressure. 1·47 + Atmosphere. 1 = 2·47 atmospheres, or 21·66 + 14·71 = 36·37 lbs.

The quantity of air that will be wanted in a cylinder can be approximately ascertained by knowing the greatest number of men that will be in the working chamber at one time, the number and nature of the lights to be used, the loss of air from the air-lock, and the constant pumping to keep the air fresh, the probable leakage of the cylinder, and the loss from the escape of air under its cutting-edge. On calculating the air required for workmen and lights, it is not prudent to allow for that purpose alone less than an air-delivering capacity of about three times that amount; but the consequences of any temporary failure of the air-supply must be considered. The quantity of air required to keep that in a cylinder in a comparatively pure state, will depend upon the nature of the soil through which the column passes, the purity of the river water, and the system of lighting used. In some situations, in very large cylinders, an equal number of, if not more, compressors, may be required to keep the air fresh than for leakage only.

It is advisable that the air be compressed equal to the full head of water. In soils which do not allow easy percolation of water, sometimes the pressure required to exclude it is less than the calculated force, the water being held back by the soil, but, as this may be only for a short time, it is not safe to work without the calculated pressure. For instance, it is recorded that a sudden increase of water pressure led to a disastrous accident at the Rheinpreussen Mine by bursting the airlock, hence the importance of the air-pressure balancing the calculated water-pressure, and no deduction being made for any diminishing influence caused by capillary attraction or a retentive stratum. As soon as the cylinder has been sunk, and there is no danger of it floating, the air-lock floor door can be opened, and the compressors set to work. These should be kept constantly pumping air in until the water is forced out, when excavating operations can be commenced on dry land. If the cylinder is situated in a tidal river, and it is well made and tight, the loss of the head of water with the falling tide may counterbalance the leakage without any pumping in of air. When the tide begins to turn, the pumps must be set to work again. After a few days' experience, the amount of air required will be ascertained; at the same time it should not be forgotten that there is always a probability of the fixed standard of air pressure not being maintained. With a declining pressure, the atmosphere in a cylinder will become misty and foggy; but with a rising pressure it will clear; if a considerable excess of air is pumped in, the atmosphere will be materially cleared; but for reasons previously mentioned, this cannot always be done.

When the cylinder enters a water and air-tight stratum, such as clay, the variations of level of the river will have no effect on the internal air-pressure, provided there is no leakage of water along the surface of the cylinder. The pressure of the air should not be raised too much above the pressure from the head of water on the stratum, or it may be percolated or injuriously affected. Sometimes springs are met with in sinking cylinders of large diameter; the air-pressure required under such circumstances, it has been found, will vary to an important extent.

The pressure of the compressed air on the air-lock floor and roof is considerable, taking a cylinder 10 ft. in diameter, and a pressure in it above the ordinary atmospheric pressure of, for instance, 36·36 − 14·7 = 21·66 lbs. = 50 ft. head of water, the upward strain on the air-lock would be nearly 110 tons. It is advisable to test the floors and roof to at least twice the pressure they will have to sustain in practice; and should they be thought weak they can be loaded to counteract the uplifting strain. The larger the cylinder, the less it is affected by sudden rises or falls in the air-pressures, as the space occupied by the air is so great that the loss or increase of a little air is not so perceptible as in a smaller column.

If a constant and increasing supply of compressed air has to be provided, the better plan appears to be to use a number of small air-compressors, particularly if the cylinder is of large diameter, in preference to one or two machines, so that if any get out of order they can be repaired without very appreciably lowering the pressure. These small compressors can all lead into one main air-pipe, and be provided with valves, so that each can be shut off at any time from the main. The main air-pipe for large works generally passes into an intermediate reservoir or receiver, which sometimes is a boiler, and then by other flexible pipes of rubber or pliable material to the cylinder. In some recent examples of air-apparatus for large works, each engine driving the air-compressing machines had its own boiler, and they were all so connected that the stoppage of one boiler or engine did not affect the rest. Precautions should be taken that the air-pipes do not foul any sharp substance that may tear or injure them. All air-hose should be of the best material, and should be tested before being used with a considerably greater pressure than that it will have to sustain in regular work. Frequently the engines and air-pumps are in duplicate, both sets being ready for work at a moment's notice, although but one set is in constant use; so that should one apparatus break down or need repairing it can be stopped, and the other set at work without delay. The air-compressing machinery is generally placed near one of the abutments on the most convenient side of the river, and the sheds for the engines, the air-compressors, pumps, smithies, repairing shops, dynamos, stores and offices are there erected. Sometimes a semi-fixed engine of from 8 to 10 nom.h.p. is used for the shallow depths, and perhaps two 25 to 30 nom.h.p. for the greater depths, such as 80 ft., to drive the air-compressors, but, of course, the engine-power required varies according to the depth of water, kind of soil, area of cylinder, and other circumstances. The system of having two semi-portable engines of say 15 horse-power instead of one, say, of twice that horse-power, to supply the power necessary to compress the required quantity of air is frequently preferred, each engine being entirely independent of the other, the air-pipes communicating separately with the air-lock.

Elaborate and heavy air-compressing machinery is not to be desired, but simple apparatus of moderate weight and size, combining efficient working with comparative cheapness, always remembering, however, that lightness and small bulk may perhaps only be obtained at the expense of economy in the production of the necessary power to compress the air. The vertical and angular system of air-compressors seems generally to be not so effective as the direct-acting steam power engine, or that constructed on the principle of direct straight line compression, *i.e.*, one in which the steam and air cylinders are fixed on the same horizontal line, and the piston rods attached to a crank working on a fly wheel. In some of the latest air-compressors, the air is first compressed to a comparatively low pressure, about 1 atmosphere above the ordinary atmospheric pressure, or, say 29·40 lbs. per square inch above a vacuum; it is then passed through an intercooler, and further compressed as desired. The great point is to reduce the strain on the machinery as much as possible, but, as in cylinder bridge pier sinking by means of compressed air a pressure exceeding 55 lbs. per square inch above a vacuum is not required to be maintained, and seldom so great a pressure exerted, it generally being from 30 lbs. to 50 lbs. per square inch, the pressure required is very much below that necessary in the case of air-compressors for tunnel work or other general purposes; but that is no reason for using old or much worn air-pumps which will probably repeatedly fail, and consequently be dangerous to employ, as the air supply may be suddenly reduced, and operations will necessarily be both slow and expensive.

It is claimed that in air-compressors the single-acting is better than the double-acting air-cylinder system, because the air is but once compressed at every revolution, and that it is therefore kept cooler as there is more time for the heat to be evolved. Unless the water for cooling is introduced into the air-compressing cylinder in the *form of spray*, as in Dr. Colladon's compressed air cooling arrangement, it is found to be ineffective as a cooler of the air during the process of compression, and unless it effects that object, it is better not introduced, the point being to cool the air *during* compression. Compressed air cannot be produced without generating heat, and the efficiency of an air-compressor is therefore reduced, but this loss is diminished by a cooling arrangement to a very small percentage of the theoretical power; however, the thermal loss must be considered with the loss by friction of the engine, as the former may be lessened by an increase of the latter.

It is necessary to cool the compressed air so as to maintain it at as little above 60° to 70° F. as can economically be effected, for the increase of temperature of air at 60° F., it being taken at the ordinary atmospheric pressure of 14·71 lbs. per square inch above a vacuum, when compressed to 2·50 total atmospheres, is no less than 158° F., or the production of a temperature of 218° F. The free or atmospheric air should be cold and moist when admitted to a compressor, a low initial temperature being economical, as it not only reduces the rise of temperature and requires the air to be less cooled during the process of compression, but less power is necessary to compress moist than dry air. It is unadvisable to keep air-compressors at a temperature below about 40° F. The object of any cooling arrangement is to take up the heat generated during compression, or as much of it as possible. This can be effected by blowing spray, at ordinary temperature, into the air-compressing cylinder during the process of compression; but if the compressed air is used expansively, the injected fine spray at ordinary temperature is employed for another purpose, namely, to prevent the air approaching too closely that of a freezing temperature, and encumbering the valves, pipes, and other parts of the machinery.

Taking into consideration size and weight of apparatus, etc., a reasonably high speed and short stroke appears to be better adapted for air-compressing machinery for use in the compressed-air system of bridge-pier sinking than the slow speed and long stroke.

One objection against hydraulic air-compressors is that the cylinders wear quickly, and, therefore, become leaky, and require to be rebored; another is that only one side of a large body of air comes in contact with the water, whereas in the spray system *diffusion and equal* cooling is attained, but there are staunch advocates of both arrangements. The air-pumps are sometimes immersed in a cistern, with a constant flow of cold water round them, to cool the compressed air. It has been found that air compressed in contact with water, and then discharged into a reservoir, leaves the machine at a temperature but very little above that of the water at any pressure likely to be required in sinking bridge-pier cylinders; also that by maintaining the temperature of the air constant during the operation of compressing it, a saving is effected in the amount of the work required for compressing and storing the air, ranging from 20 to 25 per cent. Dr. Colladon's pulverised water-compressors, *i.e.*, by injecting spray into the air-cylinder during the process of compression, were used at the St. Gothard Tunnel with so much success that in compressing air to 8 atmospheres the increase of temperature did not exceed 27° F., whereas the rise of temperature without any cooling would have been about 430° F. A condensing vessel is sometimes used to precipitate the moisture in the compressed air, in order to deliver the latter in a dry state, and the air is also cooled in the air-pumps by the injection of a fine spray of water into the cylinder with every stroke of the pump.

The temperature of the water inside a cylinder will be greater than that of the river outside; the greater the depth the higher the temperature, other conditions being alike.

CHAPTER XII.

Effects of Compressed Air on Men.

A PRESSURE of about 2 atmospheres does not appear to injure men if in health, but it depends on their temperament; those of a plethoric constitution suffering the most. Above the pressure previously indicated it is injurious to them. As the pressure of the air is increased above 2 atmospheres, the working hours of the men must be reduced; about a four-hours' shift for a pressure not exceeding 2 atmospheres, decreasing to one-hour relays for a pressure of 3 atmospheres, is usual. Many men work with comfort if the length of the shifts is shortened. Men have remained under a pressure of $2\frac{1}{2}$ atmospheres for ten hours, but this is an exceptional time. At the St. Louis Bridge, under a pressure of a little more than 3 atmospheres, several men died, or were paralysed; and the working hours had to be reduced to one per diem. It is recorded that at the Alexander II. Bridge, over the Neva, where the air-pressure in the caissons was $2\frac{1}{2}$ atmospheres, corresponding to about 85 ft. depth of water, the workmen had three-hours' shifts, and yet suffered considerably from weakness and pains in the legs and arms.

When the pressure is very considerable it is advisable to reduce the working hours, for it is believed the chief cause of paralysis in men employed in highly-compressed air is the length of time they work in it, and not more than two hours' continuous labour should be allowed at the pressures required at depths above about 85 to 90 ft. However, under favourable circumstances of clean and pure soil, and where the strength of the experienced men is *not* required to be *constantly or much exerted*, at such a depth of water as about 25 to 30 ft., men have frequently worked in compressed air in shifts of eight hours each, but when the pressure exceeds about $2\frac{1}{2}$ atmospheres, it has been found necessary to reduce the working time to about six hours; generally, if the pressure is more than $2\frac{1}{2}$ atmospheres, it is necessary to very considerably lessen the duration of the working hours.

In the winter, to prevent congestion of the lungs, owing to the sudden change of temperature on coming out of the cylinder into the air-lock, steam-coils or other means should be employed to warm the air. At the East River Bridge, the difference of temperature between the working-chamber and the air-lock was 40° F., the former being 80°, and the latter 40°. Workmen should not be allowed to go suddenly from the air-lock into the open air, especially if the pressure has been high; about one minute's rest per atmosphere is now usually allowed.

In all foundations where the plenum process is adopted there is risk to both life and limb, depending greatly upon the precautions taken, therefore duplicate or numerous safety-valves, pressure-gauges, alarm whistles, and preventive measures against fire, explosions from lighting apparatus, and accidents of all kinds, should be taken, not only to ensure the safety of the men and to give them confidence while at work, but also on the ground of true economy. To prevent the mistakes which occasionally occur when line signals are used with divers, an inexpensive speaking-apparatus has been introduced by Mr. Gorman, so that *vivâ voce* communication can be obtained with a diver. It is claimed that it is less costly than the telephone, and having no battery is much less liable to get out of order, and can be applied to any form of diving-helmet. It has been used with success at depths as great as 120 ft.

Cooling the compressed air is an important operation, which has previously been briefly referred to, as the high temperature developed when air is compressed makes it most trying for workmen. To obtain a fair average amount of work from any man, he should obviously not be placed in a heated or vitiated atmosphere, or in such a position that he is not free to move his limbs.

The air can generally be kept pure while the cylinder is sinking through permeable or porous soil, but when it is penetrating an impervious stratum the atmosphere in it may quickly become foul, and it may also be in the same condition when the bottom is covered by the hearting. Means must be at once taken to remedy this; a method that has been adopted, when, after the bottom of the cylinder was overspread with concrete, the air became foul, was by inserting through the centre of the hearting a small tube down to the permeable soil which formed the base, the upper end of the pipe being always above the hearting, the foul air thus passed through it to the bottom forced by the compressed air. Diverse opinions are held as to the cause of the pain and paralysis to which some men are subject when working under a high pressure, but it seems that with each breath, the quantity of oxygen inhaled is proportionate to the pressure, and that the inhalations per minute are voluntarily reduced nearly in the proportion between the pressure of the normal state of the atmosphere and that of the compressed air. Workmen who have been affected by compressed air, it has been noticed, are very nervous upon entering the atmosphere.

Medical practitioners prohibit violent exertion, such as climbing ladders, and severe work. Only men in good health, and of temperate and regular habits, should be allowed to work in air compressed to more than about 1½ atmospheres.

At the St. Louis Bridge, where the foundations were 100 ft. in depth, the bad effect of the compressed air upon the men was mostly felt after ascending the staircase of the shaft. A lift was therefore provided, and it was made compulsory on the men to be raised by it; they were only allowed to work two shifts per day of forty-five minutes each; they were made to lie down in a hospital boat, and were given small doses of stimulants for a short time after leaving off work.

Helmet-divers can work at a depth of 150 ft., but only for a short period, the length of the working hours extending as the depth becomes less. Depths from 80 to 100 ft. are the safe limits for most men. The working hours for divers are about the same as those for men under the compressed-air system, three quarters of an hour to one hour for great depths, such as from 100 to 150 ft., and four or five hours for small and medium depths. Native Indian divers have, without a diving dress, or any aid beyond a guide-chain, picked up tackle, etc., at depths of from 45 to 50 ft. In muddy water the matter held in suspension prevents the light penetrating, and the divers seeing; and unless the air-pump is on a fixed staging, and the ladder and air-hose protected, it is not safe to work in rough weather.

The length of time a diver can remain submerged depends principally upon the health of the man, the depth below water at which he has to work, the temperature of the air and water, their purity, and the apparatus used.

Great care is necessary in diving operations to prevent the air-hose fouling any sharp substance that may tear or injure it. The air-pipe should always be of the best material, and before being used should be tested and carefully examined, and it should never be put to work without testing after being in store. In some diving apparatus a certain amount of vacuum must be produced by the lungs to open the valve supplying fresh air; this is a drawback, and tends to prevent the divers working easily and long and with effect. Experiments have shown that if the lungs be filled with compressed air, a healthy man can easily remain under water from three to four minutes without any apparatus. A greater pressure from the air-pump is necessary with the pneumatic system of sinking cylinders than with divers at the same depth, owing to the loss of air, principally through the bottom of the cylinder and the air-lock. Divers are useful for clearing the ground of loose stones and *débris*, and for inspecting the cause of obstruction in sinking a cylinder by dredging, and for levelling and removing pieces of rock, etc.

Notwithstanding the deleterious effects of *highly* compressed air on men, it has been noticed that a beneficial action has been produced when they work at *moderate* pressures, not only in their general health, but also in the chest in particular, because of the increased quantity of oxygen inhaled under pressure ; and it has been said certain pulmonary diseases have been so cured. Such curative baths have been used in various hygienic establishments for many years.

CHAPTER XIII.

Air-Locks.

In the compressed-air system an air-lock or chamber must be constructed in the cylinder for the entrance and exit of men and materials, without allowing the egress of the compressed air in the working chamber or shaft. The maximum working pressure allowed to be used should be conspicuously indicated in large white indelible letters of enamelled iron or some substance that cannot be easily erased, both in the air-lock, gradual-pressure room, if there be one, working-chamber, and in such other suitable places as may be convenient and advisable, and the dates when the different pressures are applied should be carefully recorded ; and it is well to indicate the day of first application of the air-pressure in a prominent place to give confidence to the men. In addition to an air-lock, sometimes there is a room with two doors, the first communicating with the outer air and opening inwards, the other opening into the air-lock. It is also fitted with two cocks, with an index-finger and plate, so that workmen may ascertain the pressures. The air-lock in this arrangement is always filled with compressed air. A workman wishing to go into it enters the gradual-pressure room through the door, which he then closes, shuts the discharging cock and opens the other, and allows the pressure to increase as he feels able to bear it. When the pressure is equal to that in the air-lock he opens the door, passes into the air-lock, and descends the shaft by a ladder or staircase to the working-chamber. The operation is reversed on exit from the cylinder.

A light air-lock is made by having everything but the doors, and shoots if any, of wrought iron, and it is to be preferred to cast iron as being a more reliable material. Every precaution should be taken against bursting. The doors are liable to be especially strained, and therefore should be strengthened and supported by bar, angle or T irons riveted on all round the frame and door to prevent distortion. The height of an air-lock should not be less than 6½ or 7 ft., and there does not appear to be any advantage in making it more than 8 or 9 ft. The doors of the air-lock should be interlocked to prevent accidents, and to ensure that the entrance door cannot be opened until the door leading to the descending shaft or steps is properly closed.

Economy of the compressed air is gained by having the air-lock sufficiently large to allow all of the men forming one shift to enter at one locking. It should also be made, if possible, of sufficient extent to contain the whole quantity of material taken out by the men during one relay, so that the air-lock only requires to be emptied or drawn upon at the end of each shift. This expedites the work, and saves the men from frequent changes of pressure. Bull's-eyes of glass, for light, are often inserted in air-locks, but, owing to their being covered with dirt, very little natural light penetrates through them. Reflectors are also employed. The doors are made to open inwards, so that the internal air-pressure tends to keep them closed. The floor of the air-lock usually consists of a wrought-iron plate with a man-hole cut in it, it being firmly bolted to the cylinder, the flanges of which should be faced in a lathe and packed with approved packing. The man-hole door in the air-lock floor is sometimes fitted with an indiarubber washer, and should open downwards. If simply for the passage of men, it need not be above 2 ft. 6 in. in diameter; if excavation and materials are to be passed through it must be larger according to circumstances. The air supply and equalising-pipes pass into the air-lock, and usually a pipe for discharging any water which may percolate into the working-chamber through any sudden lowering of the air-pressure. If brass or copper pipes are used, 4 in. in diameter has been named as a prudent limit of size, and that the working pressure should not exceed about one-sixth that of the bursting pressure. Double air-locks have been used, containing one large and one small compartment, the larger for the workmen to pass, and the smaller of sufficient size to contain a bag, basket, or skip, and the necessary raising and lowering apparatus. If the pressure of air is considerable, the air-lock can be gradually loaded, so as to relieve the strain on the cylinder.

At the Argenteuil Bridge the air-lock had two diameters; the outer was 10 ft. 6 in., the inner, 4 ft. 7 in. The larger enclosed space was divided into two by a partition. One compartment was put into communication with the interior, and was thus filled with the excavated material, while the other was being emptied by the outer door, so that the loss of air in locking was diminished without interruption to the work.

At the St. Louis Bridge the caissons had a circular open-air shaft 10 ft. in diameter, which was continued to within 3 or 4 ft. of the lowest part of the compressed-air or working-chamber, and it had a spiral staircase. At its base there was an iron door, which opened into the air-lock placed within the compressed-air or working-chamber. On the air-lock entrance door being shut and the pressure equalised, the men could descend to the working-chamber almost by one step, the distance being only a few feet although the caisson was sunk to a depth of about 125 ft. below high water. By locating the air-lock *within* the compressed-air or working-chamber and at the bottom of the open-air shaft, Capt. J. B. Eads claimed that it was much the most convenient place for it ; and no extra exertion was required to reach the base or ascend the shaft, descent or ascent being in the open air ; the shaft also had not to be made air-tight, the air-lock and roof of the working-chamber alone having to be so constructed, and those in and out of the working-chamber were brought in comparatively close contact, which facilitated the supply of tools and materials and the carrying out of instructions. However, only in the largest cylinders can such an arrangement be adopted, especially if part of the permanent hearting of the cylinder is utilised as kentledge, although it can be in nearly all caissons.

A method of discharging material through a delivery pipe in an air-lock frequently used may be thus described. A discharging pipe, with closing flaps at each end, is placed through the side of the air-lock, it being inclined outwards at a sufficient angle to shoot the excavation. The process of discharging the soil is effected in the following manner. The outlet-flap of the pipe is shut, and the pipe is filled ; the inlet flap is then securely closed, and the outlet-flap opened, the material will then discharge itself. Unless the air-lock is sufficiently large to contain all the earth excavated during one shift, this method has advantages over discharging material in the air-lock and opening the door for purposes of delivery. Sometimes in addition to the discharging tubes there are, in very large cylinders, tubes for shooting the concrete for the hearting into the air-lock. The inclination of these latter pipes should be the reverse of the excavation tubes. A code of signalling must be arranged between the workmen in the air-lock and the men outside. Signalling by means of an acoustic tube and vibratory diaphragm has been employed with partial success, but is generally abandoned owing to the noise made by workmen rendering it difficult to understand the signals. Whistling signals have also succeeded, the compressed air being allowed to escape through sonorous reeds. Electric signals are the best. The telephone has been used, but, owing to noise, it cannot under such circumstances be considered thoroughly reliable.

At the Boom Bridge, over the Rupel on the Antwerp-Tournai Railway, where the excavated material from the interior of the cylinder was discharged through the outer air-lock door, the inner end of the spout opening inwards and the outer door of the pipe necessarily outwards, it is obvious that if, by mistake between the men in the air-lock and those outside, the outer door was opened at the wrong time, the flow of air would be very dangerous, and perhaps fatal to the men in the air-lock. An arrangement was therefore devised by which the safety of the men was secured at trifling expense. It consisted in locking the fastening bolt of the outside door by means of a sliding pin, which was worked by a rod passing through a stuffing-box into the compressed-air chamber, the pin being withdrawn only by the men inside the chamber, and not until they had previously closed the door on the inner end of the spout.

A discharging tube, consisting of buckets formed with india-rubber lips, working in a perfectly true and smooth cylinder, has been employed to save leakage of air by dispensing with the method of discharging materials from the cylinder into the air-lock, and then outside; but the leakage of air was so great that it had to be abandoned. A frequent rule, when the air-lock is also used as a spoil-lock, and will contain all the material excavated during one relay, is for the men to cease operations in the working-chamber half an hour or so before the shift terminates, in order to remove the soil previously accumulated in the air-lock.

In deciding upon the position of the air-lock in a cylinder, space should be economised, the amount of air wasted should be caused to be a minimum, and the safety of the men should be secured in case of an accident to the cylinder. If an air-lock is placed in the cylinder below the water-level outside it may be dangerous, especially if there is a considerable range of tide, and also an uneven river-bed. On the other hand, among the disadvantages of the upper air-lock system, the air-lock must be sometimes taken off and replaced, and the air-shaft must be ascended by the workmen when under pressure, a not unimportant question if the air is compressed more than 2 atmospheres, for every endeavour should be made to avoid unprofitable exertion of the men at all pressures. If the air-lock be placed above the working-chamber, although at the bottom of the air-shaft, it must be entered from the top, and left through the bottom. Side doors cannot be used, but they can be if the air-lock is within the working-chamber. Owing to the great difference in area of the chamber of a cylinder and that of a caisson, ranging generally from 1 to 100 to 1 to 200, these advantages and conveniences of access are of but little importance in the cylinder, but very great in the caisson. As a matter of prudence, if the air-lock is close to the working-chamber, or inside it, or below the water-level, it is desirable to have an additional refuge or safety room. For cylinders, on the whole, it would appear that it is preferable to place the air-lock on the top, but in the case of caissons of considerable area, or of cylinders of very large diameter, it is more conveniently situated at the bottom. It is always advisable to provide for capping the cylinder, or shaft of a caisson, so that, if necessary, it can be made a receptacle for compressed air.

CHAPTER XIV.

Working-Chamber, and Method of Lighting it.

THE height of the working-chamber, or chamber of excavation, should not be less than 6 ft. 6 in., and from 7 to 8 feet is a preferable height. Each man requires about an area of from 20 to 25 sq. ft. to enable him to work freely. It is seldom that more than 8 to 10 men can simultaneously and profitably work in a cylinder of moderately large area. To crowd the men is to waste labour. The chief aim should be to equally excavate the material so as to prevent tilting of the cylinder, and to manipulate the excavation so that the resistance of the ground and the pressures are equal over the whole area of the cutting-ring, and any outside local looseness of soil obviated. It is well to have the working-chamber painted white, in order to obtain the greatest possible amount of reflected light, and it must be thoroughly stayed in all directions by angle-irons and gusset-plates.

When the air-lock is at the base of a cylinder, the air-supply pipes should project about 3 ft. into the working-chamber, so that in the event of an accident, and water rushing in as quickly as the air was forced out, which would air-seal the bottom of the supply-pipe, the space between the end of the air-supply tube and the roof of the working-chamber would contain a layer of compressed air, so that the men would not, in that case, necessarily be drowned; but it is questionable whether many men would, under such circumstances, be sufficiently calm to avail themselves of the refuge. The working-chamber sometimes alone contains the compressed air, the air-chamber being fixed below its ceiling.

Shafts for lowering the materials to the air-chamber, when the latter is at the base of the cylinder, are generally arranged as follows:—A tube or pipe, about 2 ft. in diameter, is fixed from the top of the air-chamber to the summit of the cylinder, with doors at the top and bottom, the lower opening into the air-chamber. When the upper door is open, the lower is held in position by the pressure of the air in the working-chamber, and by ordinary arrangements. The supply-shaft is then nearly filled with material, or with as much as is desired, which being effected, the upper door is drawn up, compressed air is sent into the pipe, and when the pressure in it is equal to that of the working-chamber, the air-chamber is signalled, the fastenings of the working-chamber door are removed, and the material is deposited. Many accidents, however, have arisen through mistakes in the signals, and an automatic arrangement from above is preferable ; but notice must, of course, be given the men below that the material is about to be deposited, in order that they may keep away from the mouth of the supply-shaft. To ascertain whether all the material put in the shaft at one time has been discharged at the bottom, a rod, or other means, should be employed. A thoroughly trustworthy foreman should see that the top door is always shut, and that the requisite amount of compressed air is let into the supply-shaft before the discharge door is opened, or the shaft will be blown out, the compressed air in the working-chamber will be set free, the water will immediately flow in, and the men in the air-chamber will probably be drowned, all ordinary lights will be extinguished, and the moisture being, by the sudden absence of pressure, set free from the compressed air, would cause a mist, and in addition there would be the roar of the escaping air, which would render it almost impossible for men to grope their way to the ladder. It is advisable to conspicuously mark the ladder, whether by a phosphorescent plate, luminous paint, or by other means, so that its position can be ascertained in the dark.

The supply-shaft at the lower end should be gently splayed, and the bottom door and fastenings must be made sufficiently strong to sustain the weight of materials in it, and in filling the shaft it should be ascertained that the material does not get jammed, or it may have a sudden fall, and fracture the lower door. At low tide, owing to the decrease of the hydrostatic head, there will be less chance of water getting into the working-chamber than at high water. Any small agitation of the water on the surface of the ground in the working-chamber will permit the air to escape if the undulations allow the water to get momentarily below the edge of the cylinder, therefore it should be kept as still as possible.

A strong light in the working-chamber is a necessity, not only to penetrate the mists that prevail from time to time, but to illuminate the whole internal base of the cylinder, in order that the excavation may properly and equally proceed. At a pressure of two atmospheres above the ordinary atmospheric pressure the wick of a candle will rekindle when the flame has been blown out, therefore inflammable materials should be kept from the vicinity of the lights in the working-chamber of a cylinder. In a fire in the East River Bridge caisson, as soon as the water-pipes were stopped playing upon the timber, it would re-ignite. Candles produce much smoke, owing to their rapid, but incomplete, combustion under an excess of air-pressure, and they are liable to be extinguished by air-currents. The nuisance of smoke has been overcome by reducing the size of the wick and the candle, and by mixing alum with the tallow, and steeping the wick in vinegar. Candles have been burnt in closed glass lamps, the air being brought from the surface. Lamps are of but slight use, as they smoke more than candles, and the oil, to a certain extent, is dangerous. The relative volume of oxygen consumed should be considered in determining the kind of light to adopt.

If gas is used for illuminating a cylinder, it is necessary to have its pressure always 1 lb. or 2 lbs. above the air-pressure in the column; the pipes should therefore be of extra strength, so as to obviate the possibility of their breaking, which would probably cause an explosion. A gas pump is sometimes used for obtaining the necessary pressure. At the St. Louis Bridge the gas tanks were filled with water from an artificial reservoir having a head of water always slightly in excess of the caisson pressure. Into these tanks the gas was discharged from small cylinders under a pressure of 225 lbs. The immediate effect was to force the water from them back into the reservoir until the tank was full, when the supply was stopped. The pipes leading to the caisson remained opened, and the gas passed through them under the pressure due to the artificial head of water. By means of glass gauges the contents of the tank could be watched to be replenished as often as necessary. The gas tank was placed below, in the air-chamber, so as not to require building up as the caisson sank. If the gas is pumped directly into the tank, the stroke of the pump creates an unpleasant jumping of the flame. As danger will arise from leakage of the pipes, and from leaving any cocks open, the lighting should be carefully supervised. The sense of smell under compressed air is greatly lessened, and leakage is not easy to detect.

At the St. Louis Bridge the gas burners kept the temperature below at 80° to 85° F. Gas was found to cost one-fifth of the calcium or limelight, and about one-third that of candles; it, however, produced a considerable amount of heat, and vitiated the air more than candles. A candle when blown out was instantly relighted for twenty times successively, and a woollen garment quickly ignited if brought momentarily in contact with a flame. Candles with fine wicks had only 5 per cent. increased consumption at a pressure of 46 lbs. per square inch, but a cotton wick in alcohol no less than 200 per cent. at the same pressure. The alcohol wick flame, instead of being blue, changed to a white colour, giving three-fourths as much light as a coach candle. The relative cost of candles as compared with alcohol was as 1 to 2. In the East River Bridge caisson, after reaching about 20 lbs. pressure per square inch, the gas lights smoked very

badly; the cause of the smoke was deemed to be a lack of ventilation of the flame, or circulation of air around it, the size of the burners was therefore reduced as the pressure increased, with the result that there was but little smoke, less gas burned, and a better light. Mr. F. Collingwood, in a paper read before the Lyceum of Natural History, U.S.A., stated that from numerous experiments on the burning of stearin candles when in compressed air, he found that "the amount of consumption at various pressures is approximately as the square roots of those pressures," and a waste of one-third took place from flaring of the flame while in motion. The above rule shows the number of candles that will be required at any depth, after the quantity wanted has been determined at any other depth. General experience has demonstrated that the electric light is the most suitable for the working-chamber of cylinders, and that it should be used in preference to any other yet devised, as, when properly arranged, it has invariably given satisfactory results. The portable lamps are most useful for this purpose; however, great care should be exercised to prevent the men being placed in darkness from any cause, as accidents may then arise, and it is therefore advisable to have a light or lights continually burning entirely independent of the electric illumination. In caissons it has been found that a few arc-lights are not so suitable as a considerable number of small 16-candle glow-lamps placed around the caisson, chiefly because the height of the working-chamber is insufficient to allow of the effective diffusion of light. The impediments that have been experienced in adopting the electric light in caissons and cylinders have been chiefly confined to the difficulty of preserving the insulation of the wires, and keeping the lamps free from dirt and moisture.

CHAPTER XV.

EXCAVATING AND DREDGING APPARATUS FOR REMOVING THE EARTH FROM THE INTERIOR OF A CYLINDER OR WELL

It may be said that the excavating or dredging apparatus has to perform the most important part in cylinder sinking, for without an efficient means of removing the earth from the interior of a cylinder the latter cannot be sunk to the required depth; it is therefore a matter of much importance to employ the best machinery for the soil that has to be excavated and raised, as each kind of earth requires a cutting and disintegrating apparatus that has been specially designed for it in order that it may be completely successful, and the point to determine is, what is the best machine to use under the particular circumstances. Some of the advantages and disadvantages of using the compressed-air system for sinking cylinders have been referred to in a previous chapter. A few of the advantages of sinking cylinders by means of dredger apparatus are now given :—

1. The hours of labour need not be reduced and are not regulated by the depth of the foundations below water-level.

2. There is no danger to the men, and no liability of sudden and fatal accidents occurring.

3. The comparatively small cost of the dredging apparatus.

4. No air-lock and working-chamber are required, and the cylinder need not necessarily be made air-tight.

5. With the exception of the dredging apparatus and lifting machinery no other special plant is required.

6. Less skilled labour is necessary.

7. Provided the dredging apparatus is adapted for the earth to be excavated and raised, it is independent of the depth of the foundation below water-level.

8. The cost of working does not increase according to the depth, for a dredger-excavator can be efficiently employed at any ordinary depth with but little additional expense, that being principally due to more time being occupied in raising and lowering the apparatus and consequently to the fewer lifts that can be made.

9. Its portability and easy erection.

Some of the disadvantages may be considered to be :—

1. That as the foundations cannot be inspected in the open air when the excavation is completed, and only by divers, or by means of a diving-bell, it cannot be known whether or not the whole area of the base is equally supporting the hearting of the cylinder.

2. That it is by no means easy to excavate the soil equally over the whole internal area of a cylinder, and when the action of the grabs or buckets can only be in the same perpendicular line, the soil may not be sufficiently loose to fall equally around a central hole made by the excavator, consequently, should the ground not be of the same character, the cylinder may become inclined. Means will, however, be named by which this may generally be prevented.

3. Unless the interior of the cylinder is so arranged that nothing can be caught by any projections, the grabs or buckets may be held, and the dredger have to be broken or abandoned from this cause, but it is in great measure preventable.

4. The difficulty of excavating close to the cutting edge, especially in cohesive soils such as clay, and sufficiently near to it to cause the earth

to fall into the central hole, or become loose enough to be excavated and raised by the dredger-grabs or buckets.

5. The tediousness and difficulty of removing unexpected obstructions such as tree-stumps, large boulders, or masses of conglomerate, and the then perhaps necessary employment of the compressed-air system, either by a diver, diving-bell, or by an air-lock, etc., so as to disintegrate the obstruction sufficiently to enable the excavation to be raised.

6. The difficulty of excavating cohesive soils by dredgers, also clayey silt, and compact sand and gravel, which latter, however, have seldom to be excavated except in seams. NOTE.—Some means will be named by which this difficulty may be much lessened and perhaps entirely avoided.

7. The heavy strain on the hoisting apparatus and wear and tear of the buckets and grabs, consequently the latter especially should have as few parts as practicable, and those that come in contact with the soil should be additionally strong.

Most of the principal advantages and disadvantages of dredger machinery for excavating and raising the earth in the internal area of a cylinder having been named, reference is made to some of the chief points to be especially considered in grab or bucket-dredger machinery to be used in cylinder sinking.

1. It should excavate the earth over the whole internal area of a cylinder, or nearly so, and not be dependent upon the soil around a central hole falling into it.

2. The grab or bucket should easily enter the ground, and sufficiently to cause it, when closed, to be full of earth, and it should shut tightly and readily, either pushing in or out any boulders or lumps of material.

3. Very little or no earth should be washed away or drop out during the operation of raising or hoisting a bucket or grab through the water.

4. As little water as possible should be raised to the surface with the excavation.

5. The grab or bucket should readily discharge its contents, and not require to be cleared.

6. It should be simple in construction, with as few parts as possible, be not liable to get out of order, be easily repaired, and occupy a comparatively small space.

7. Special provision should be made for extra strength in any closing chains and in the bucket edges.

8. Any grab or dredger designed to excavate the earth under the cutting ring should be capable of doing so under its whole area, so that no lumps remain to be removed by divers, if they will not fall into the central excavation pit, or any hollows that may be formed between them.

9. Preferably, no special lowering or lifting apparatus should be necessary; but it is well to remember that such a quality may only be obtained by a sacrifice of efficiency.

10. The arms or bent levers, which, on being moved, cause the grab to excavate and hold the material, should not when opened project much beyond the grab edges, in order that the scoops or grabs may penetrate and excavate nearly the whole horizontal area over which they extend on being lowered.

11. It should not require the constant removal of heavy plant when a ring of the cylinder has to be added.

12. It should so perform the excavation that, as nearly as practicable, only the net cubical contents of the subterranean portion of a cylinder have to be removed; and it should not disturb the surrounding earth or cause "blows" or "runs" of soil, and so probably prevent vertical sinking.

These may be stated to be the chief requirements, but there are others that have to be considered, and they will be named. Here it is not intended to describe in detail or illustrate the various machines that have been introduced for the purpose of excavating the earth in a bridge-cylinder well or caisson, as most of them have been illustrated and described in the various engineering journals and the Minutes of Proceedings of the Institution of Civil Engineers and other scientific societies, but to comment upon some features to which attention should be directed in almost all such machines and the soils for which they are considered to be especially adapted. It would be most difficult to say which is *the* best grab or bucket-dredger. Some are more suitable for one kind of earth than another, and for comparatively little depths. For considerable depths it would appear that those actuated by a strong central rod are to be preferred to those having looser means of opening and closing; and those which occupy, when fixed for descending, the least area and have the simplest and most direct-acting parts should have the preference.

From a study and analysis of many cases in which various kinds of buckets and grab-dredging machinery have been used in bridge cylinders and well foundations, almost all have been successful when applied under the circumstances for which they were intended to be used; the chief difficulties to be overcome are those of penetration of the scoops and equal excavation over the whole internal area of a cylinder or well. When the material to be excavated and raised is loose soil, there are many kinds of most efficient bucket-dredgers actuated by chains, rods, bent levers, etc., the scoops acting on stationary pivots or by means of other devices having one combined object, namely: easy descent, penetration, gathering up, perfect closing, and gentle raising of the excavated material without allowing any earth to fall over the sides of the bucket or grab. For considerable depths, and in cohesive or hard compact soil the most certain plan of action, and one that will seldom fail, appears to be to first sufficiently disintegrate the earth by a separate apparatus, so that it can be expeditiously gathered by a bucket or grab, rather than to proceed by attempting to excavate, collect, and raise the material by one machine at one operation, which may be ineffectual unless divers can be sent down to loosen the earth ready for the grab or bucket to lay hold of and raise it. By first separating the soil into sufficiently small pieces so as to be readily gathered and raised to the surface by the grabs or buckets, any time occupied by the first operation will be soon compensated by increased speed and certainty of action, and by the dredger being full or nearly so when it is lifted, instead of, as frequently is the case, only partly filled. For removing boulders, cohesive or hard compact soil, the ordinary dredger-buckets have too much surface to readily penetrate the earth, and may be unable to do so, and a dredger is required that will plough the soil. Sand also under a considerable head of water may be difficult to penetrate with ordinary scoops, and in clayey or sandy silt the scoop may not bite or enter it sufficiently to cause the bucket to gather its proper quantity of soil, for it then often merely scrapes the surface, its powers of penetration being insufficient to enable it to grasp the earth.

Apparatus which may fail when unassisted, if aided by heavy jumpers and sharp cutters will often remove earth of the usual description met with in cylinder sinking, but generally simple quadrant bucket-dredgers are ineffective in clay, tenacious, or moderately compact soils. If it is found that the dredgers will not make their own holes, or enter the ground, which they may not do in silt, sand having boulders in it, or in clay, cutters or jumpers can be used. However, the boulders may be too large to be moved by a dredger and too hard to be broken sufficiently small by machinery working from a height in water; it may then be necessary to adopt the compressed-air method of sinking; but divers should be tried first, although excavating by means of helmet-divers is not economical. On the Continent the tendency of late years has been to abandon the dredger-system and adopt that of compressed air; but there is no reason, except the requirement that the foundations must be laid dry, so often decreed in Continental specifications, and unless serious obstacles are expected to be encountered in sinking, why it should be renounced in favour of the compressed-air method, as great improvements have lately been introduced in dredger-excavating machinery. If obstructions, such as *débris* and large boulders, or other obstacles, which cannot be readily broken by helmet-divers, are not likely to be met with in sinking a cylinder, and if the column cannot be readily kept dry, excavating the soil by machinery under water is the cheapest method to adopt.

When the compressed-air system is used in sinking, or the excavation effected by divers, ordinary excavating tools can be employed, but no expense should be spared to procure the best, most efficient, and expeditious tools that can be obtained, as any extra expense thereby incurred will be quickly saved by the work being accelerated. Consequent upon the short hours men can work in compressed air, or in a diving' dress, every effort should be made to economise their labour, as the actual working time may be as little as one-thirteenth of the usual hours under ordinary circumstances, and the wages considerably higher.

The means of lowering, closing, and raising bucket and grab-dredgers have been well considered, and it is in the direction of increased efficiency of the cutting and breaking-up apparatus so as to feed the buckets and grabs, and cause them to become quickly and easily filled, and the earth excavated over the entire internal area of the cylinder to prevent any additional cutting of the earth from underneath the bottom ring, that the greatest scope for improvement exists; but, as has been before stated, experience appears to point to the advisability of an effective first use of the cutter and jumper tool to disintegrate the ground, and a grab or bucket-dredger to raise the loosened soil, rather than to attempt to effect too much with the grabs or buckets and so court failure in cohesive soils and those difficult to penetrate; whereas by a combination of the two systems success will be almost certain. It is impossible to be sure that logs, boulders, or tree trunks will not be met in such variable soil as the beds of rivers, therefore in cylinder sinking it is an advantage to have an apparatus ready in a few minutes to break up any such obstruction should the work of the grabs or buckets become unsatisfactory; however, if the trunk of a tree or a boulder be encountered, probably the quickest and most effective plan is to send down a diver to direct the cutters or jumpers, and so shatter the boulder, or by sawing, axing, chipping, barring, and by chains being placed round the trunk, to pull it into the cylinder, and so enable it to be raised.

Almost every kind of earth requires a specially shaped tool, grab, or bucket, and the suitability of the form and capacity of the dredging apparatus causes it to be successful or to fail, and entirely different results will be obtained when these points are carefully considered. As it is seldom certain that no hard or tenacious soil will occur in sinking a cylinder, it is important in selecting a dredger-excavator for such work that it be adapted to excavate and raise any such stratum, and the question should always be initially determined whether a cutter or jumper shall be employed simply for breaking up the soil and a dredger for raising the material. When a cylinder has to be sunk to a moderate depth and there is every probability of the soil being comparatively loose, a bucket or grab-dredger may be sufficient; but if the depth to be sunk is considerable, say more than about 40 ft., and a hard stratum is expected to have to be excavated, experience seems to indicate that it is better to disintegrate the hard material independently of a dredger, and only use that for collecting and raising the loosened earth, as then the soil can be easily penetrated, and operations are likely to be successfully accomplished without delay, while attempting to thrust a bucket, scoop, grab, or spade-dredger through tenacious clay or hard soil may not only be ineffectual but result in breaking the apparatus, for pronged spades are liable to be bent, turned up, and broken; therefore, as a precaution against injuring the forks, it is advisable to first disintegrate the soil with jumpers or cutters, or to send down divers to effect that operation.

When the force with which a dredger can be dropped into the soil is simply that of its own weight falling through a certain distance, it is obvious, bearing in mind the extent of the cutting edges of the bucket, that a sharp-pointed jumper or chisel having a penetrating area of, say, less than 1 sq. in. must have a greater power of penetration than that of a bucket blade, whether serrated or not, having a continuous flat or inclined edge, which, although pointed, has a thickness of $\frac{3}{8}$ or $\frac{1}{2}$ an inch and a length generally exceeding 18 in.; and if a small boulder should happen to get under the cutting edge the apparatus is likely to tilt and become ineffectual.

The defects of dredgers for undercutting the curb are that they do not do so *equally under the whole area* of the curb or cutting edge, consequently, short pieces of earth remain between those portions excavated, which induce "blowing" or "running" of the soil. These lumps of earth have either to be excavated by divers, to be undermined, or left unsupported in such a way that the central excavating hole can be deepened sufficiently to cause the remaining pieces to slip in, a system which is not conducive to either economical, quick, regular, or vertical sinking. This is one of the chief difficulties with scoop-dredgers, as many can only with certainty be lowered in and near the centre of a cylinder or well, consequently in any cohesive soil a hole like an inverted cone is excavated, and as the sides do not readily fall into it, unless the hole is filled by other means, the quantity dredged at each lift is very small, and progress necessarily slow and uncertain.

It not unfrequently happens, even when the dredger appliances are so successful as to equally excavate the material over almost the whole internal area of the cylinder, that it refuses to sink notwithstanding additional weighting, and that it is necessary to excavate under the cutting ring of the cylinder or permanent ring of the hearting used as kentledge, or the curb, if the well system is used. When the undercutting apparatus fails, divers must be sent down to disintegrate the material under the cutting ring or curb, and cause it to fall into the central hole. In clay, if the dredger apparatus only breaks up the earth for a portion of the area and leaves a wall of 2 ft. or 2 ft. 6 in. in thickness, it will generally not fall into the excavated central hole unless it is disturbed or separated, and therefore the cylinder will not continue, to sink. In either a cylinder bridge-pier or caisson any internal staging or timbering should be so arranged that the excavation can extend to the edges of the rings, so that it is not necessary to employ divers to shovel the earth towards the centre in order that the dredgers may gather and raise it. An instance may here be mentioned of the difficulties caused by the excavation being effected over but a small portion of the area of a caisson. At the Poughkeepsie Bridge, U.S.A., the foundations of which are 124 ft. below high water, and were reached when mud, clay, and sand had been excavated, and rest upon strong gravel overlying rock, an open grillage of cribwork was used for getting them in, it being divided into pockets or cells from which the material was excavated by dredging. The cribs were 104 ft. in height, and the top finished 20 ft. below water. The cribwork was built on the shore and towed out. It was divided into weighting and dredging pockets. Fourteen dredger-cells were simultaneously worked, but their area only amounted to one-fourth of that of the crib, therefore considerable masses of earth were left under the cutting edges, and consequently the wells or dredge-cells were often carried 30 ft. below the base before the crib would sink, and the sinking was irregular; sometimes the crib went down suddenly 10 ft., and did not then descend vertically.

With respect to buckets or grab-dredgers, cases have occurred in which it was found that although the buckets were suitable for loose silty soil, they were too large and blunt for pure sand, although it could be easily excavated ; and also that in compact and viscous silt the form that is effective in loose soft silt is unsuitable, it being necessary that the bucket edge be more pointed so as to enter the earth and not merely scrape it, and also discharge more easily the soil on the bucket being tipped.

Opinions are somewhat divided as to the relative merits of dredgers of large and small capacity ; however, in a large dredger the weight of the excavator is less in proportion to that of the material raised than in the case of a small bucket or grab, thus in a very large excavator its weight may be as little as 0·75 of that of the earth lifted, whereas in the small dredgers it may vary from 1·2 to 1·7 time the weight of the earth, therefore in proportion much more dead weight has to be raised each time, provided the large dredgers are always full, which point has been previously referred to in this chapter. A dredger of small capacity can excavate at almost any point in a cylinder, and is generally raised full, whereas large dredgers are liable to be nearly empty. On the other hand, in small cylinders, their capacity being as little as 2 to 3 cub. ft. instead of $\frac{1}{2}$ or $\frac{3}{4}$ of a cubic yard or more, an additional number of lifts have to be made, but the large machine can only act near the centre of a cylinder, and when a "run" of soil occurs, the hole dredged by it being generally much below the level of the inflowing soil, the apparatus becomes buried, considerable delay is caused, and perhaps the machine is broken. Under similar circumstances the small dredger can be readily abandoned or pulled up as desired without interfering with the working of other similar dredgers in the cylinder.

It is usually a serious matter when large grabs are caught in a cylinder, for work is then entirely suspended; but by having no projections or abrupt internal bends or splays, these accidents may be avoided to a considerable extent, therefore a variation in the size of the interior of a cylinder is a disadvantage in using dredger machinery. When the dredging apparatus is light, and fresh rings are added to the cylinder, no heavy hoisting machinery has to be removed and replaced; however, if there is independent staging, the lifting apparatus can be so arranged as to allow of fresh lengths of the cylinder or well being added without affecting the machinery for raising the grabs or buckets. Perhaps in cylinders of small diameter the best plan is to have two sizes of dredger-grabs or buckets, the smaller holding from as little as 2 to 4 cub. ft., and the larger 7 or 8 cub. ft. or more, according as the size of the cylinder will permit, duly taking into consideration the depth below water, the size of the cylinder, the nature of the soil to be excavated, and all other circumstances. In a 6 ft. in diameter free working space, a 7 to 8 cub. ft. capacity of the dredger, grab, or bucket will probably be as large as can be conveniently worked, its dimensions being increased as the available open area allows of easy operation. Experience seems to indicate that a larger capacity than about $\frac{3}{4}$ of a cubic yard is of doubtful economy and efficiency in *cohesive* soil, because of the difficulty of penetration, in brief, the capacity of the buckets will be governed by the diameter of the free working space in the cylinder, and the nature of the soil, but the bucket and grabs are usually made to contain from $\frac{1}{2}$ to 2 tons weight of excavated material, and most frequently between $\frac{1}{2}$ to 1 ton; and at moderate depths as many as fifteen grabs may be made in one hour, but Messrs. Bruce & Batho's dredgers have successfully worked with a capacity of 5 cub. yards. Much depends upon the lifting power available, for the larger the capacity of the dredger and the more cohesive the earth to be excavated, so will the power of the lifting apparatus require to be increased. The excess of power advisable to employ may be as much as three times the weight of the excavation to be lifted, for the weight of the dredger has to be raised, and the excavator has to pull out the ground with it, so the force required, in addition to friction, etc., is to some extent dependent upon the tenacity of the earth. Although the full bucket capacity of material may be occasionally raised in a dredger, from one cause or another it is not prudent in cylinder sinking to rely upon more than about half the full contents as the result realised in *continuous* work performed by ordinary labour.

Some hard and stiff material, provided it will break in lumps, can be excavated by dredger-grabs when the apparatus used is specially designed for that purpose, but in tenacious clay and concreted or solid gravel, *i.e.*, with the particles joined by a cementing substance, or in very compact firm sand, it is advisable to first separate the soil with cutters or jumpers, and this may be absolutely necessary. Most bucket-dredgers work well in loose sand, but when clay and compact soil are encountered, they are usually not so successful. However, it is generally agreed that the shape of the dredger causes the success or failure of the apparatus. Simple bucket or quadrant-dredgers have been incapable of excavating a soil, but when a prong was added, and the cutting edge of the bucket was of a suitable shape for penetration, it would excavate the ground. An instance may be here referred to. In order to prevent a grab-dredger refusing to excavate below a certain depth, and to ensure that a sufficient distance in the ground was penetrated to cause the earth to fill the grab when it was closed ready for raising, Mr. W. Matthews, M.Inst.C.E., had a few prongs about 1 ft. in length riveted on the outside of a grab-bucket, with the successful result that it descended to any required depth, the prongs loosened the ground, and the bucket collected and raised the earth as desired.

Although grab-dredgers do not excavate quite so evenly as bucket-dredgers, it is seldom they so unequally do so as to cause inconvenience. In boulder or pebbly soil, consisting of loose stones and silt, sand, or sandy gravel, stones are liable to become fixed between the quadrants, and also between the teeth of the grabs, and then the smaller material falls out, and although the excavating operation is successful the lifting is not so, as much of the material is dropped. Grab-dredgers are not designed to be driven into the soil by a ram falling upon them, but, as their name implies, to grab or lay hold of the earth on being lowered. The best grab-dredgers are so framed as to penetrate the soil on the lifting chain being drawn up, even when the grab is gently lowered upon the ground, its weight being sufficient to cause it to enter the earth. The single bucket and clam-shell dredgers succeed in loose material having separate particles, but they are not adapted for firm clean sand or hard tenacious earth, or clayey silt, or soil that possesses some tenacity and viscosity although comparatively liquid.

In considering the dredger-bucket or grab, its size when opened for descending, its power of penetration and excavating, capacity, easy raising, and also the unaided discharge of the lifted material have to be considered. Buckets that will freely discharge clay and viscous earth of all kinds are to be preferred, if not lessening the general efficiency of the apparatus, to those requiring a man with a bar to free the soil from the raised bucket. Experience has shown that large buckets discharge the contents more freely than those of smaller capacity, and that the bucket should taper horizontally and vertically according to the nature of the material to be discharged, and *not* have parallel sides. In a case where there was great difficulty in discharging the material from the bucket, the backs were made movable so as to throw out the soil. The shape of the bucket when ready for raising should be such that it will be filled entirely with earth and its contents be raised without loss, and holes in the buckets should allow of the escape of any water so that as little as possible is lifted with the earth.

In bucket-dredgers or excavators the wear of the closing-chains and the edges of the buckets is considerable. All working parts should therefore be of the strongest material and steel be used in preference to iron. In the best dredger-excavators all holes are bushed, and the pins and bushes made of case-hardened steel, the arms, sliding collars, and the plates, etc., being of cast steel. Some special features of bucket and grab-dredging apparatus may be thus described:—The buckets are made of double-riveted steel plates having heavy replaceable steel jaw plates. Clay-grabs consist of bent prongs, and form a combined bucket and grab, in which the prongs fit in between each other ; their upper part is an ordinary close bucket, the idea being to enclose the looser material in the top portion of the bucket. As the clay becomes more tenacious the distance between the prongs is increased, there being as many as twelve or so in a combined bucket and grab for excavating and raising loose and soft material, nine or ten for shingle and earth that can be raised in lumps, but for hard clay soil only four or five prongs in the same length of grab-bucket. In fine sand, difficult to penetrate, the prongs or tines are fixed tightly together, or the bucket has a serrated edge. For boulders and hard lumpy soils the prongs are strengthened by a bar riveted on each side at a distance of about two-thirds of their length measured from their points.

In bucket-dredgers, it would appear that when the blades of the buckets are of spear or V form, there is considerably less chance of their tilting than when they are simple quadrants, but buckets of segmental form are much steadied by outside prongs being fixed as previously described. Some of the most important and practical improvements in dredger-excavators are those of Messrs. Bruce & Batho. 1. The introduction of buckets shaped so that they present a pointed cutting blade, and form a hemispherical bucket when closed, which may be familiarly described as half an orange, instead of a semi-circular bottom with two flat sides.

Thus four pointed blades are in simultaneous operation to disintegrate the soil instead of two flat quadrant bucket edges. 2. One pair of blades being larger than the other, they excavate a greater area than the other two, consequently the excavator does not fit into the hole previously made. These are important improvements, as the power of penetration is much increased and the central hole difficulty is prevented, two of the principal drawbacks in dredger-excavators for cylinder-bridge piers.

A slight change in the nature of the earth to be excavated may make all the difference in the working of the dredging apparatus, for instance grab-buckets have succeeded in excavating and raising hard quartz sand, which had round grains, but were not so successful when the sand had flat and pointed grains, as the particles became wedged and were consequently not easily penetrated or separated in their natural position. For excavating close to the cylinder rings in soft soil, and at small depths, a bag and spoon-dredger is very useful, as it cannot injure the rings or become jammed, and if it breaks, it is not a serious obstruction. It can either be worked by hand or by a small steam engine on the staging.

A question to be considered is how many cubic yards of earth can be taken out of the cylinder in a working day. The quantity of soil to be removed is generally much more than the contents of the subterranean portion of the cylinder, and according to the excess of material so will be the disturbance of the outside earth, which will usually assume the shape of an inverted cone, the base being at the level of the bed of the river. In soft soil, in addition to the expense of excavating the excess of material, there is the danger of the cylinder being for some time feebly supported laterally, until the earth has become consolidated, which, however, may be prevented by scour; therefore, care should be taken to preserve the external bed of the earth in order to prevent "runs" of soil in the cylinder, and to reduce the required excavation as near as possible to that of the cubical contents of the subterranean portion of the column. When obstructions, such as logs and trunks of trees, have occurred in *loose* soil as much as three to four times the contents of the cylinder have been necessarily excavated, but this may be considered an extreme case. Means have been named by which the outside surface disturbance of the river-bed may be lessened or prevented, and a method that might be frequently tried with little expenditure is that of hardening suitable soil by injecting into it liquid Portland cement. The area of the base that can be excavated by a dredger should be compared with the total area of the cylinder, so as to ascertain the width and mass of material left around and under the cutting edge by the excavator, for the extent of this strip will influence the speed of the excavating operations, and the smaller it is, the quicker the removal of the earth and *vice versâ*. It is advisable to have a large and small dredger or some means of excavating close to, if not under, the cutting ring in case of necessity.

With regard to the hoisting apparatus, it should be effectively worked at any depth, in order that there may be no limit to its operation. To mention in detail the hoisting apparatus that might be used would cause reference to almost every known kind of lifting machine, and it will here suffice to state that it should be simple ; should quickly raise the dredger apparatus at an even speed and without jerking, or some of the contents may be lost, and the dredgers be raised partly empty ; it should have sufficient power to pull up the grab or bucket when it penetrates the earth, so that it may be completely filled ; it should be as light as possible consistent with the required strength, for it has to be raised at each lift ; it should be removable ; and, if it can be so arranged, not necessarily fixed on any staging that rests upon or is supported by the cylinder rings, so that on fresh rings being added it need not be removed ; and it should not in working have any tendency to pull the cylinder towards the source of power, and so perhaps be the cause of the column not sinking vertically. Derrick-poles fixed on staging, entirely independent of the cylinder, have therefore been used so as to comply with the last two conditions, the excavation being raised by a steam hoist.

In some examples of river bridge foundations sunk by compressed air, a tube, projecting beneath the cutting-ring of the column, has been inserted in the cylinder, an adjustable dredging-machine working in it, the bottom being sealed by the water, and the earth being raised by the dredger. The material to be removed by dredging-machinery in such a tube must be pushed under the lower edge of the shaft into the pool of water underneath it, which must always be maintained, in order that the dredger may be properly fed and no air escape. At the Forth Bridge the hoisting of the material was done by a steam engine fixed outside the air-lock, and working a shaft upon which there was a drum inside the air-lock. By means of a stuffing-box passing through the air-lock roof, there was no escape of the compressed air.

At the St. Louis Bridge, to facilitate the excavation in the caisson, an extra tube was inserted in the centre, and down to the level of the bottom of the air or working-chamber. The water in the tube was at the same level as in the river, and in the pipe an endless chain, with dredger-scoops attached, rotated round pulleys at the top and bottom of the tube. Thus the sand was raised without the escape of air from the chamber, or passing the material through the air-locks. The men in the working-chamber shovelled the material to the bottom of the tube, which was in the water, where the dredger-scoops took it and discharged it at the top of the caisson. Somewhat similar apparatus was proposed by Mr. Wright in 1852 to be used at the Rochester Bridge foundations, Kent.

At Chicago, dredging apparatus was fixed upon a traveller on the top of the cylinder, and upon the platform of the traveller was a carriage for removing the material; the frame-work being adjustable, so that the sand could be dredged to a depth of 6 ft. below the bottom of the cylinder. The excavating-machinery consisted of two endless chains, on which were placed fourteen iron buckets, each of about a capacity of ¾ of a cubic yard. The buckets were driven by a small portable engine fixed on a traveller. This brickwork-in-cement cylinder was 31 ft. 6 in. in internal diameter, and was sunk through 20 ft. of quicksand down to solid clay.

The discharging spouts of dredging-machinery have been arranged in the following manner:—The buckets delivered the sand into a movable spout worked on a cam, so that the spout was brought forward to receive the contents of a full bucket as it mounted the top; the spout was then drawn back, so that the empty bucket passed down clear. At the new Prague-Smichow Bridge, the excavated material was emptied from the dredgers into a shallow tray, having a lateral automatic motion communicated to it by an arrangement of cranks and levers, and was thence discharged into chambers, of which there was one on either side of the air-lock, fitted with inlet and outlet sliding doors, from which it was again removed, the outlet door being closed when the inlet was opened, and *vice versâ*. These doors or valves were actuated by vertical rods passing upwards through stuffing-boxes and worked by manual power from a platform above and outside the air-lock. At the bridge over the Ticino, at Sesto Calende, Italy, the sandy clay excavation in the cylinder was removed in wrought-iron buckets holding 0·39 of a cubic yard each by a small three-cylinder engine, worked by compressed air, placed above the air-lock, the material being discharged through a pipe having suitable doors.

CHAPTER XVI.

Notes on some Dredging Apparatus used in Sinking Bridge Cylinders and Wells.

WITH respect to the different apparatus generally used for removing the internal earth from a bridge cylinder or well, a few notes of a practical character are given under the head of each machine.

BAG AND SPOON DREDGER.—This is very useful for mud, soft and loose soil, if comparatively small quantities of excavation have to be removed, and for clearing trenches for cofferdams, the site of piers, the corners of caissons, or in any situation where a narrow trench has to be dredged in such soil, and also for excavating close to the cutting-ring of a cylinder. It may be used where grab or bucket-dredgers may be either impracticable or unsuitable. If used in firmer soil than silt and mud, instead of the ordinary bag, forked spades can be fixed to the end of the long handle. The excavation at the Victoria Bridge, London, Brighton and South Coast Railway, was conducted in the following manner. The bed of the river was levelled by bag and spoon dredgers before the cylinder rings were pitched, then the excavation in the column was effected by the same means until the clay was reached, when the water was pumped out, and the excavation carried on in the open air. At the Charing Cross Bridge, South Eastern Railway, the bed of the river over the site was levelled by dredging, a sufficient height of rings of the cylinder were then pitched so that the top reached above the water level, a bag and spoon dredger was used inside the cylinder, and the mud and gravel so excavated until London clay was reached; the water was then pumped out, and sinking conducted by ordinary open excavation. The bag and spoon dredger was found to be a better system to use, in many ways, than excavating by helmet divers.

INDIAN JHAM, OR HOE-SCOOP WITH HANDLE.—This machine has been very largely used in India, and is occasionally employed in loose sand for small depths. For dense sand the scoop is made of thin plate-iron, so as to penetrate the ground. It is not well adapted for depths exceeding about 25 ft. The process is slow, and only economical at depths at which the bag and spoon, the earliest system of dredging, is applicable. At the bridge over the Ems at Weener, the excavation was done by steam-dredger, hand-dredger, and Indian scoop. This last, 2 ft. 1 in. by 1 ft. 7 in. by 7 in. in dimensions, was attached by a hinge to a vertical guiding rod, and was dropped, edge first, into the sand to be dredged, being pressed down by a man standing upon a step attached to the rod. It was then wound up by a crab. By 140 lifts from a depth of 13 ft., it excavated 9 cub. yards a day, the cylinder being 13 ft. 1 in. in external diameter, and sinking 1 ft. 8 in. Although the steam dredger was somewhat cheaper in its work, the scoop was preferred because it excavated round the edges of the cylinder so much better, and the columns consequently sank more evenly than when the dredger was used. In the large cylinders of 18 ft. 4 in. in external diameter, a steam-dredger and two scoops were worked together, excavating 15¾ cub. yards per day.

STEEL POINTED RAMMER FOR ROCK OR CONCRETED SOIL.—The steel pointed ram system of excavating rock under water, introduced by Mr.

Lobnitz, of Renfrew, in which in soft rock the points of the rock-cutting rams become automatically, through work, at an angle of $\setminus 70°/$ and in hard rock $\setminus 43°/$, and remain at those angles, has been used with success on the Suez Canal for breaking up rock for dredging. The rods were placed in a frame and allowed to drop about 18 ft. They weighed four tons each and were about 40 ft. in length, and were hoisted by a chain and let fall. This method might be used in cylinder sinking to break up a hard stratum, or an obstruction, as its object is to dispense with blasting, and to shatter the rock, etc., under water without inspection of the surface.

STONE'S CHISELLED-RAIL JUMPER, used on the Delhi Railway.—It was composed of rails chiselled at the lower end to a point, fished together to make any required lengths. It was raised by a crab engine, or by manual power, to a height of 8 to 10 ft. and dropped, and was used in stiff clay with great success for separating the ground in masses weighing a quarter to three-quarters of a hundredweight ready for the bucket-dredgers to raise the loosened soil. It was found that it possessed the great advantage of enabling the earth to be broken up over the whole area of the cylinder or well, and the jumper could be guyed and made by tackle to fall in any place.

Generally, the solid bar or rail jumper, pointed at one end, is sufficient for shattering any soil likely to have to be removed in cylinder sinking, except when a thin stratum of a very hard nature is encountered; more powerful means may then be required.

STRONG'S CEYLON GOVERNMENT RAILWAY EXCAVATOR.—This consists of a cylinder 4 ft. in diameter and 2 ft. in depth. The bottom is divided into 6 parts of equal length fitting on and within a circular plate, the latter being 5 in. in width. Six equi-distant sharp-pointed bent picks are attached by pins to bars actuated by a central rod. The apparatus is heavily weighted and lowered so that the sharp points pierce the clay, and when they are drawn up the material is raised. It is only intended for clay and tenacious soil that can be raised in lumps.

GATWELL'S EXCAVATOR.—Grab-dredgers in boulder soil, that is, those which in closing form a kind of bucket thus U, have sometimes failed ; for instance, at the bridge over the Sutlej, Indus Valley State Railway, Col. Peile, R.E., the engineer-in-chief, stated that at a depth of 35 to 45 ft. a band of compact, intractable clayey silt was encountered which resisted various ordinary dredgers and sand pumps. Chisels and cutters of various forms, applied to the ends of heavy bars, were used to cut up the surface, and divers were kept continually at work, but no progress was obtained in the sinking for many weeks. Then Gatwell's excavating apparatus was employed. Two blades were used for excavating the earth in the cylinder, and once they brought up 30 cub. ft. of clay at one lift; and one for cutting under the bottom ring or curb. It may be described as a combination of the Indian jham and the ordinary dredger. It consists of two pointed scoops which penetrate the ground on being lowered in a vertical position. On touching the ground a hook becomes disengaged, and by hauling chains the blades are drawn apart, and, in excavating the soil, assume a horizontal position under the earth excavated, the nearly flat scoops being then raised. Prongs are also attached to the apparatus for breaking up clay, and a side excavator is used for under-cutting the curb. This latter consists of a kind of spade bucket fixed to a long rail lowered vertically by a chain, a rope being attached to it below the hanging chain, it is then pulled over to such an angle that it will undercut the curb, and raised. The Gatwell excavator does not close inwards, like the ordinary grab and bucket-dredgers, but forms a figure when being raised resembling two trays ↓↓ ↓↓ , thus it cuts from the centre outwards, the material being on the ledges A A, which are slightly rounded, and no boulders can prevent it raising the material. An advantage in working with this apparatus in cylinders in silty soil is that as the shelves turn outwards and are level when ready for raising, no silt or loose sand can run out owing to the buckets not being closed tightly. They occupy slightly more room when opened for lowering than the inwardly-raising bucket-dredgers. These excavators have succeeded where some ordinary inwardly-closing quadrant-dredgers have failed in clay soils, and are generally successful.

BRUCE AND BATHO'S EXCAVATORS.—The special features are that they are circular on plan and hemispherical in form, having three or four or so V-shaped pointed blades, and therefore can be used nearly of the same size as the cylinder, and so excavate close to the cutting edge or curb, thus avoiding any undercutting. They have been employed with a central pole or tube, which forced the blades into the soil, applied at the end of the jib of a crane which held the apparatus in the centre of the cylinder in a fixed position, and have been used with success at a depth of 130 ft. below water level. They are especially designed for penetrating hard and tenacious material, and are usually worked by a chain in soft earth, and by a spear in hard soil.

FOURACRE'S DREDGER AND SPIDER CLAY-CUTTER.—The former has been used successfully in India for dredging sand and mud. It consists of two segmental dredger-buckets hinged to a cross-head and worked by lifting chains. The spider clay-cutter for loosening clay and other soil so that it can be removed by dredging, consists of six picks hinged at equal distances apart to a central shaft; each pick has a connecting rod hinged to a boss placed about 3 ft. 6 in. above the hinged end of the six picks. On a pile-driving ram, made to slide in the central shaft, descending, the boss is driven down with the connecting rods, the picks are thus forced into the soil and break it up sufficiently for raising. The six picks are extracted from the clay by pulling a chain attached to the boss, and are so lifted.

MILROY'S EXCAVATOR.—This apparatus, one of the earliest designed, has been largely employed; it usually has eight ordinary spades for mud, sand, and loose permeable soils, pronged spades being used for gravelly earth, marly and ordinary clays. It has excavated clay to a sufficient depth for the water in the cylinder to be pumped out. A 5 ft. in diameter apparatus, 2 ft. 3 in. in depth, has raised 1¾ cub. yard at one lift. At the Caledonian Railway Viaduct over the River Clyde at Glasgow, where the greatest depth of water was 23 ft., and the cylinders were 15 ft. in diameter, the soil penetrated was principally sand, clay, and mud, with a few small beds of gravel; the foundations being 85 ft. below high water. A steel digger was used, weighing 3 tons 16 cwt., arranged somewhat like the Milroy apparatus. It had 12 blades, each heavily weighted on the back to assist penetration, and was 8 ft. in diameter, 15 in. in depth at the sides, and 22 in. in the centre. The blades were attached to the centre of the digger by a system of bars similar to the framework of an umbrella, and were drawn up by four chains. The blades, suspended from a monkey hook, were sunk in a vertical position, so that the moment they struck the bottom, the hook was detached; they were then drawn together and the machine raised. The digger sank into the ground by its own weight. When raised, the original hook was attached to the blades, which, on being lifted, fell apart and deposited the earth in a waggon. In favourable soil the average lift occupied five minutes, and 1½ cub. yard was raised. The digger frequently brought up boulders weighing from 7 to 8 cwt. each.

IVES'S EXCAVATOR.—Especially introduced for excavating in stiff clay as well as all less cohesive soils. It is driven vertically into the earth by a pile-driving monkey. The lower portion is hinged and forked with six or more prongs, which penetrate the ground on the hinge-catch being withdrawn, the prongs are pulled round till they are at right angles to the vertical rod, and then the whole apparatus is raised. Instead of a pronged fork, a scoop is used for loose soil such as sand. In reconstructing some bridges on the Delhi Railway, Mr. Charles Stone, M.Inst.C.E., stated it was found a perpendicular position was necessary to give effect to the driving power, and it was somewhat difficult to keep it vertical, and that the apparatus could not be lowered to work at any great depth except in the centre of a cylinder, but this comment applies to most excavators.

Among simple devices used for loosening clay may be mentioned that employed at the Ravi Bridge on the Punjab Northern State Railway, which consisted of a screw 9 in. in diameter fixed to a 2¼ in. gas tube.

This was repeatedly screwed into the soil, and was successful in loosening clay so that it could be raised by the dredger-buckets.

At the Katzura Bridge, near Kioto, the foundation was compact gravel and sand, and could not be raised by ordinary excavators. DIACK'S EXCAVATOR was used. It consisted in a flange rail being hammered to a point, and two iron bars of the same width as the web of the rail being fixed to it by three bolts, and bent so as to form a quadrant of a circle like a fishing net, the projecting length being about 2 ft., and the width 1 ft. 4 in.; the bottom of one ring had a cutting edge, and to the other five teeth were attached to loosen the gravel. Two canvas bags were fastened by cords looped through holes in the bar. The required length of shaft was obtained by joining lengths of rails by fish plates. The apparatus having been lowered into position was worked round by manpower being applied to a lever attached to the rail shaft. About $\frac{1}{2}$ of a cub. yard was brought up at each lift, and six to twelve lifts were made in one hour.

At the Bookree Bridge, Great Indian Peninsula Railway, when a hard conglomerate stratum of marl and gravel and impacted sand and hard clay or marl was reached, bucket-dredgers would not penetrate it. Mr. R. Riddell therefore employed a long vertical iron rake, the shaft being made of old rails bolted together, pointed at the end, which sunk into the ground at the centre of the cylinder. About 3 ft. above the bottom a cross-piece armed with strong steel prongs was bolted, and the rake was revolved by the capstan bars from the top. This apparatus loosened the hard soil and enabled the dredger to lift the excavation. There was no difficulty after the boulders, *débris*, and conglomerate had been so disintegrated. When large stones were met, they were shattered by divers with hammer and bar, and then raised in the bucket-dredgers.

At the Rokugo river-bridge, Japan, Mr. R. V. Boyle, C.S.I., M.Inst.C.E., has stated that Bull's hand-dredger was first used and answered well in brick cylinders, 12 ft. in diameter and 2 ft. in thickness, in fine sand near the surface, but in coarse gravel and soft mud the efficiency of the hand-dredger became much reduced. Kennard's improved sand pump was used and did good service in the coarse gravel but was not found suitable for dealing with the mud. A double bag excavator was then tried. It consisted of two bags, each being fastened to a frame of iron, the lower part of which formed a cutting edge. These frames were fixed on opposite sides of a vertical bar, by which they were made to rotate and dredge a circular hole. In deep bridge-wells, the two frames were bolted to a square intermediate socket fitted loosely on a vertical rod, which remained suspended in the cylinder while the frame and bags were lifted and emptied at the surface.

Any required pressure upon the bottom to make the cutting edges effective was obtained by loading the frame with weights slipped into a cross-bar attached to it. Eight men were able to turn the excavator by bearing on tillers keyed to the vertical rod, and under favourable circumstances, four to six lifts were made in an hour, and 5 or 6 cub. ft. of mud were raised in the smaller apparatus, and 12 to 16 cub. ft. in the larger dredger.

Many of the grab-dredgers designed for excavating clay or tenacious soil, follow the principle of ordinary bucket-dredgers, the chief alteration being that the buckets instead of being made of curved plates are formed entirely of ribs or prongs having pointed ends, the lowering and raising apparatus being almost identical. When clay, boulders, hard or tenacious soils, such as compact gravel veins, have to be excavated, ordinary bucket-dredgers are not adapted for effecting the excavation. Briefly, simple bucket-dredgers are useful in loose earth, grab-bucket-dredgers for moderately hard soil readily penetrated, and especially designed dredger-excavators, such as those that have been described, for clay and compact earth; or the cutter and jumper system of first separating the ground, the bucket-dredgers being simply used for gathering and raising the loosened soil.

SIMPLE BUCKET-DREDGERS.—Almost all form a nearly semicircular bucket when closed. Bucket-dredgers are so well and favourably known for excavating and gathering all kinds of loose material that there is no occasion to describe them. The one great objection to them is that they are only adapted for *loose* soils. The buckets have been raised full in certain loose deposits, but when the earth is harder to penetrate, they are only partly full or nearly empty. Much depends on the character of the earth. The varieties of sandy, silty, and clayey soils are very numerous. It is necessary to know the nature of each before deciding as to the suitability of the dredger. Impure loamy or argillaceous sand, or sand derived from soft sandstone rock, or that has round grains, will generally be readily raised ; but sand derived from quartz or hard sandstone rock, rough, angular, hard sand which is clean, clear, and translucent, and consists of fragments of the most durable rocks, will usually be difficult to penetrate ; as also clean fine sand, *i.e.*, sand almost wholly comprised of minute particles of hard rock. In silty soil it will generally be found that the *clayey* silts are difficult to penetrate by bucket-dredgers, as the edges of the buckets slide on the surface or do little more than scrape it ; whereas, if the silt is of a decidedly sandy nature it is more easily separated, and therefore penetrated. In the former case, the tendency will be for the bucket, as it were, to float on the surface, in the latter, to press the harder particles into the muddy clay or softer portion of the mass, and so penetrate it. Clayey silt is by no means a satisfactory earth to excavate and raise, its viscidity preventing both easy penetration and discharge ; again, sandy soft silt is usually raised with so much water, that although a bucket may be full, only about one-half of its contents can be called excavated material, and the greater the depth of water through which it has to be raised, the more charged with water the earth is likely to be. Small openings in the dredge-buckets will allow the water to a certain extent to drain away, but it is very difficult to separate the water from the earth without some loss of the soil. Those who have had difficulty in excavating soft silt at a considerable depth below water, say from 30 to 70 ft., have stated that it was not only advisable, but cheaper, to adopt the pneumatic system, and put men in to excavate such material in the ordinary way, especially so when the earth is very soft and the cylinders of small or moderate diameter.

In clayey soils all depends on the nature of the clay. If it will readily fissure or break into lumps, or has much sand in it, penetration will be easier than if it is comparatively pure clay of any kind. Except in loose soils it is generally advisable to serrate the edges of the buckets, but perhaps the most important recent improvements in bucket-dredgers are, (1) the introduction of the pointed hemispherical form of scoop ; (2) the attachment of prongs or bent forks to the outside of the buckets so that they are able to penetrate the earth, and can be drawn together by the closing bars or chain, and, (3) loading the buckets so as to cause them to descend with greater force. The first two improvements are the most important.

With regard to the use of a bucket or grab-dredger actuated by two chains, one for opening the bucket, or grab, and the other for closing and raising the load, they are worked by a special form of crane ; whereas the single chain bucket or grab-dredgers, *i.e.*, those having one chain for lowering, closing, and raising, can be worked by an ordinary crane ; but the latter when closed at the bottom of a cylinder, or in the earth cannot be opened until they are raised, and therefore if boulders or other obstructions are encountered, the chains may break, the dredger become fixed, or the catch hooks that release the grabs or buckets on the latter reaching the ground may not act properly through being obstructed by lumps of earth, *débris*, or loose stones.

One of the best known and most successful dredgers for bridge cylinders, it having been largely used in India, is BULL'S SIMPLE QUADRANT DREDGER. It is specially adapted for sand and loose fine sandy gravel, and possesses the advantage of having few working parts, and of being very simple. The buckets are fixed to curved arms. In lowering they are kept open by a catch, and when lowered it is released by a rope, the closing and hoisting chains pass round a pulley attached to the ends of the lower arms. There are two arms, and each extends over the opposite quadrant during lowering, thus, on their being pulled together, a considerable grip of the earth is obtained.

Among some other dredgers may be named Sir Bradford Leslie's rotary plough, or boring head, referred to in the next chapter under sand-pumps, suction, water-jet and compressed-air dredgers; Molesworth's dredger; Stoney's helical excavator; Furness & Slater's telescopic dredger, used on the Thames Embankment works at shallow depths from 18 to about 30 ft. below water, and described as suitable for sand, compact sand, and porous gravel.

When the excavation in the cylinder is completed, the interior should be cleared, and the bed levelled by a diver, and all earthy matters that may adhere to the sides and flanges should be removed. To prevent sandy soil becoming impregnated with mud or muddy water before the hearting is deposited, arrangements should be made that, as soon as the cylinder is free from *débris*, and the base properly levelled, the hearting is deposited over the entire internal area of the column, and any "runs" of earth or percolation of water prevented.

CHAPTER XVII.

SAND PUMPS, SUCTION, COMPRESSED AIR, AND WATER-JET DREDGERS.

WITH regard to sand pumps, subaqueous-dredging on the suction or water-jet system, and compressed-air dredgers as applied to excavating the earth in the interior of a cylinder or well, it is not here intended to describe in detail the various apparatus, but to refer to some few points connected with them. In the sand pump, the suction-pipe draws in water with the material, the proportions of sand or mud to water being different according to the nature of the earth. A mixture of 5 of water to 1 of sand has worked well. The cohesive properties of clay soils prevent the employment of excavator-pumps, but, when they are helped by cutters disintegrating the earth, they have been used in loose ground of that nature. Sand and fine gravel are the most suitable earths for them, but they have been employed with radiating cutters on the lower part of the movable bottom, which, being rotated sufficiently, break up seams of clay or tenacious soil for raising by pumping, *i.e.*, into pieces somewhat smaller than the suction-pipe. Their chief disadvantage is that they lift a large quantity of water with the sand, and consequently much of the power applied is wasted, and perhaps a "run" of earth is induced by a flow of water being caused. Experience has shown that it is better to work sand-pumps by bands and not by gearing, *i.e.*, by a yielding medium in preference to rigid driving, because, should the pumps become choked, which is sometimes the case, the power being as it were elastically communicated prevents injury to the machinery.

Some consider that hydraulic dredgers should preferably be worked by a centrifugal pump, because of its comparatively few working parts ; and as the action is continuous and in one direction, there is no stoppage or change of the stroke, and the material is steadily ejected and therefore cannot settle. These are important advantages. A reversal of the flow which induces settling should always be avoided as much as possible. Among the pumps especially adapted for pumping sandy water may be named the centrifugal, rotary, pulsometer, and chain. For such purposes pumps having complex or delicate parts, or pistons which fit closely, or that have other than ball or clack valves should not be used, and it is well if any pump valves and seats can be removed so that they can be inspected. The wear of the cylinders in a direct-acting pump in sandy water is very considerable, and it may be advisable to use some other kind.

Boring with hollow rods and a continuous current of water might perhaps be occasionally used to break up sand difficult to dredge by ordinary bucket-dredgers, but the current of water must be continuous or the tube and the cutter may become fixed, the object of the machine being to flush out the earth as it is excavated by the cutting-tool. The necessary velocity of the current is ascertained by the ease with which the tube penetrates, and the force-pump pressure can be so regulated as to produce the desired velocity. This method, it has been claimed, possesses advantages over that of other boring systems. For soft ground the flushing current passes down an inner line of pipe forming the boring rods, and rises to the surface through the lining pipe. On the contrary, in boulder ground, the water is admitted through the lining tube and passes out loaded with the material through the central hollow rod, the diameters being increased ; a $3\frac{1}{2}$ in. tube being considered the maximum working size. The pressure required usually varies from about three to five atmospheres for ordinary depths. The blades of the cutting-tools should be perforated so as to allow the water to circulate. When the flushing tube becomes plugged, a method of clearing is to raise it 10 or 15 ft. and pump rapidly for some time.

Some devices for breaking up the soil sufficiently small for dredger pumping will now be considered.

At the foot of the suction or dredger pipe, which is flexible to allow of its being spread around, a scraper is fixed ; when this is dragged over the bottom it loosens the material sufficiently for the earth to be drawn into the suction-pipe with the flowing water.

Sir Bradford Leslie's rotary-plough or boring-head is a combined compressed air, boring, excavating, and lifting apparatus. It was used at the Gorai River Bridge, on the Eastern Bengal Railway, in very hard clay as well as in ordinary earth and sand, the soil being discharged by a constant current of water in a pipe. The average daily sinking of a 14-ft. cylinder was 4 ft. 6 in., but 9 ft. was really done in one day, the other being occupied in removing and refixing the apparatus on an additional ring of the pier being added.

The quantity of earth raised depends upon the power applied to drive the plough and the volume of water flowing up the pipe, and is quite independent of the depth. It raised and discharged anything that could pass through the discharge pipes. The apparatus consisted of a horizontal disc-plate, with four triangular blades at right angles to each other, projecting underneath, and armed with cutters, which, when revolved, excavated a conical hole 9 ft. in diameter. The plate was bolted to an annular pipe of 13 in. inside and 26 in. outside diameter. The space between the inner and outer pipes was made air-tight by annular flange plates riveted into the ends of each 9-ft. length of pipe. The shaft therefore consisted of a vertical pipe surrounded by a series of air-jackets. The boring head was worked by a small compressed air engine. The earth was removed by a current of water constantly flowing up the pipe. For this purpose a 12-in. syphon pipe was provided, the inner leg of which was immersed in the boring-shaft, and the outer leg in the water of the river; then, by connecting the suction of the air-pumps with the syphon, the air was exhausted from it, and, being replaced by water, a flow of water from the cylinder into the river was immediately established proportionate to the quantity thrown into the cylinder by the two 13-in. centrifugal pumps.

At the Hooghly Bridge the dredging was effected by a special boring gear, one set in each of the three excavating chambers, similar to that at the Gorai Bridge; but it was driven by steam instead of by air-pressure or turbines, and the syphons through which the earth was discharged into the river were charged by Körting ejectors instead of by air-pumps.

Hutton's sand pump, and Burt & Freeman's sand and mud pump were used on the Amsterdam Ship Canal works, St. Petersburg-Cronstadt Canal, and on the Lower Danube. The latter machine had stirrers or knives on the vertical pump shaft, and by means of jets of water from a force pump impinging under great pressure on the dredged material, it was disintegrated before it entered the pump, thus the scope of this appliance has been largely extended, and clay of moderate consistency was successfully dredged in the new cutting of the Sulina branch of the Danube.

Colonel Schmidt's is another form of dredger for excavating clay as well as loose earth by the aid of cutting knives and a special arrangement of centrifugal pump, suction, telescopic suction, and discharge pipes.

At the new Tay Viaduct the soil was chiefly silty sand, with occasional beds of gravel, boulder stones, clay, and red sandstone ; it was found that the steel digger of the Milroy pattern lost a considerable quantity of the material while being hoisted through the water. This led to trials being made with various kinds of pumps in order to raise it without loss. The best results were obtained from a 12-in. centrifugal pump, the suction connections of which were thus arranged Two flexible hose-pipes, each 6 in. in diameter, and 20 ft. in length, were placed in the bottom of the cylinder, the ends being brought together and joined into one 12-in. pipe leading to the pump on the platform. A diver was then sent down who manipulated the suction pipes, so that while one 6-in. pipe threw up sand, the other kept the pump free by drawing clear water only. As much as 40 cub. yds. had been pumped up in one hour, causing a subsidence of over 2 ft. in a 23-ft. cylinder. When the tide was too low for pumping, the digger was used. In clay strata, as the material could not pass up the pump, the two flexible hose-pipes were removed, and the water was pumped down as far as possible, giving additional pressure on the bottom owing to the difference of level of the water. In this manner cylinders had been sunk as much as 11 ft. in thirty minutes, and the material was afterwards taken out with the digger.

At Dunkirk Harbour the dredging was effected by the two-fold action of streams of water injected under pressure into the sand and by exhaustion, *i.e.*, by injection and suction, for each of which operations a separate centrifugal pump was used. One pump was employed for driving the water into the air-chamber, and thence to a hydraulic injector or sand-pump, similar to that used for excavating the foundations of the St. Louis Bridge, which required only two pipes, one passing down to the sand-pump, and the other brought the materials to the surface. Thus there was no obstruction to the use of the air-locks, etc., in the caisson at that bridge. The lower end of the injector, which rested on the bottom, was made of cast iron so as to sink readily into the sand. The water was injected under pressure down one pipe, and passed out of three small tubes, which projected slightly from the casting and stirred the sand, the sand and water being drawn up a separate pipe by the action of the other centrifugal pump. Both of these pipes were flexible at their extremities. The apparatus is more suitable for comparatively large areas, as in a caisson, than for a cylinder bridge pier.

At the Alexander II. Bridge over the Neva, a Körting hydro-ejector was erected to pump up the mud, and it acted satisfactorily. The water pressure used was 10 atmospheres, and the issuing water contained 26 per cent. of solid matter, as much as 38 cub. yds. of mud being raised in one day.

Reeve's vacuum excavator was used at the first Tay and Severn bridges, etc., etc. It was considered suitable for sand, silt, mud, loose clay, and small gravel. The material was excavated by means of a flexible suction-pipe discharging it into receivers, from which the air had been exhausted. It was worked under water, the tide rising and falling within the cylinders.

At the new quay walls at Calais Harbour, where the soil is very fine and movable sand, vertical square built masonry walls on a strong concrete curb were sunk by the pressure of water. The depth of the foundation was 8 to 11 metres. The walls were 1 metre in thickness and 8 by 8 in dimensions, with an octagonal shaft 4 metres in diameter. They were sunk side by side, leaving 0·4 of a metre between them. The first, third, and fifth well were first erected of one series, then the intermediate, and lastly the 0·4 of a metre space was excavated by a water-jet, and filled with concrete, which dove-tailed the whole series together by filling two pairs of grooves which had been formed in the sides of the contiguous masonry wells. The wells were sunk by injecting water under their cutting edge by means of wrought-iron pipes carried down through a central shaft, and splayed outwards so as to direct the jet upon the sand beneath. Thus loosened, the material was brought up by a centrifugal pump, whose suction-pipe descended in the centre of the shaft and drew sand and water from the bottom of the conical cavity which was gradually formed by the disengagement of the sand around its sides. The necessary pumping machinery was mounted on wheeled trucks, which ran upon a tramway parallel to the line of the wells. For the water-jets, four Tangye pumps were used, supplied with steam from two small vertical boilers, while a centrifugal pump was driven by a separate portable engine and boiler. Twelve wrought-iron pipes were used as water-jets, and were divided into four groups; the three pipes of each group were connected with one of the pumps by flexible rubber tubes. The jets were so directed around the cutting edge as to excavate the sand regularly, and in general the verticalness of the wells was easily maintained. Sinking a well 4 to 4½ metres took twelve to fourteen hours, being equal to 20 cubic metres of excavated material per hour. In a bed of clay the operation was difficult and tedious, but succeeded in beds 1 metre in thickness. The system was also used for smaller wells 4 metres by 4 metres, but although the rate of descent was faster, it was not easy to sink the wells vertically. Some difficulty was experienced in working the centrifugal pump owing to the settlement of sand in the suction-pipe, which tended to choke the valve at the foot of the pipe, when pumping was temporarily stopped. It was remedied by attaching to the valve box one of the wrought-iron pipes, through which a jet of water was at such times delivered into the valve box just above the valve, and by means of the circulation of a continuous stream of water the deposit of the sand was prevented.

AIR-LIFT EXCAVATING APPARATUS, AND THE WATER-JET SYSTEM.—Sand, mud, and loose soil can be raised by the escape of compressed air through a discharge pipe leading into the open air at the top of the cylinder, and can be emitted in a continuous stream. Where the air space is small, as in a cylinder, this method of ejecting sand, etc., by direct force of the air may be difficult to keep in regular work; but in a caisson, because of its size, the objection vanishes. At the St. Louis bridge sand was forced up under a pressure of 10 atmospheres, or about 150 lbs. per square inch, one $3\frac{1}{2}$-in. pipe raising 20 cub. yds. per hour 125 ft. in height when continually worked. At the East River bridge, at a depth of 60 ft., sand was continuously discharged through a 3-in. pipe for thirty minutes at the rate of 1 cub. yd. in two minutes, and fourteen men, standing in a circle round the pipe, shovelling as fast as they could, were required to supply the mouth of the air-discharging pipe with sufficient material. General Smith stated that with this apparatus men need only enter the compressed-air chamber to remove an unusual obstruction, and that such an appliance is required in sinking cylinders to very great depths, and necessarily the greater the depth the more efficient the air-lift. Of course this method can only be used when the material will yield and flow with an air current, unless it is previously separated. At the Glasgow bridge over the Missouri, Kansas City, St. Louis, and Chicago Railway, some of the excavation in the caisson was removed by means of an Ead's sand-pump, but it was found that the most economical and rapid method, when sand and gravel had to be excavated, and the pressure exceeded 5 lbs. per square inch, was the air-lift, for here it simply consisted in a 4-in. pipe passing down through the roof of the caisson being provided with a valve within the air-chamber, and terminating in a short goose-neck. The sand being piled round the lower end of the pipe, and the valve being opened, the escaping air raised it with great velocity. It only required a moderately increased supply of air, which is always desirable for changing and keeping fresh that contained in the air or working chamber.

Dredgers for the removal of sand or silt by an injection of compressed air, instead of by suction, have been used successfully in soft silt, sand, and gravel. Jandin's apparatus was used at Havre and Saumur. The principle of it is that of forcing air through a number of holes in a pipe surrounding a main pipe, the compressed-air being sent into the internal raising tube by an injector. The pressure causes the water to rise in it, thus dredging the loose soil at the bottom, and lifting a mixture of water and sand, the latter being 25 to 40 per cent. of the volume. This apparatus is actuated on the same general principle as Sir Bradford Leslie's boring-head previously described. At the bridge over the Po at Casalmaggiore, Parma-Brescia Railway, the earth, which was of a sandy nature, was cast by men into a box holding about 7 to 8 cub. ft., and water was pumped through a pipe to mix with it. The mixture was then forced out by another pipe in an almost continuous stream by the pressure of the compressed air in the chamber. About 30 per cent. of sand and 70 of water were so discharged, the volume being about $5\frac{1}{4}$ cub. yds. per hour.

Sir F. Bramwell, many years ago, suggested that at the bottom of a cylinder a massive plate, sufficiently heavy that the water-pressure underneath could not raise it, should be fixed ; that a pipe reaching to the top to force water might be attached to the seal plate, and a worm wheel fitted to the pipe to cause it to rotate, or that motion be imparted by a hand spike. In brief, that the hydraulic method of sinking piles by forcing the sand *outside*, as used in disc piles sunk by the water-jet, might be applied to sinking cylinders in loose sandy soils capable of being forced *outwards* by water-pressure from the interior. Advantages claimed for this system are that no "blows" or "runs" of soil into the cylinder can take place, and therefore the outside earth is not nearly so much disturbed, a cylinder is better supported laterally, and any movement of the surrounding ground is likely to extend equally in all directions.

Experience with the water-jet system has shown that although very efficient and economical in sand, silt, mud, or soft clay, when the sand is clayey or gravelly it loses much of its efficiency, and in gravel it is not a desirable method to use. In its application to cylinder sinking the jets should be so arranged that they discharge the water under pressure in such a way as to ensure regular and vertical sinking. The reason of the ineffectualness of the water-jet in gravel and gravelly sand is that although the sand and any earthy matter are washed out, the stones of the gravel remain and accumulate until they form a barrier which the water-jet cannot remove, it forcing any sand or earthy particles through the interstices of the stones, the result being a layer similar to a pebbly beach. The water-jet is almost useless in such soil, or when large timber chips, so often found embedded in the earth near docks and piers, have to be removed.

When water is forced vertically in the direction of the axis of a pile by means of an oblique hole being made in it near its point, down which the water is injected into a hole made for a short distance along the axis of the pile, sinking is much faster than when a pipe jet is brought to the point of a pile, therefore if water-jets are used in cylinder sinking it is well to remember that the vertical action of the water is the most effective in increasing the fluidity of sandy soil ; for that is the principal effect of the water-jet, the especial object of the appliance being to produce such a state of fluidness that the pressure necessary to cause effluxion is as little as possible; thus there are two actions, and obviously the more water the greater fluidity, and consequently the less pressure required; therefore a considerable jet discharged with moderate force is more effective than a small jet emitted at a high pressure; still an excess of water is undesirable, a sufficient fluidity of the earth being all that is necessary. As the necessary fluidness of a mass of gravel, pebbles, or solid clay cannot be attained by the application of a water-jet under ordinary pressure, it is in sandy soil having fine particles easily transported under slight pressure, or in loamy soils which on water being forced into them quickly become liquid mud, that the water-jet system is particularly successful. The nozzles of the discharging jet should be properly formed in accordance with the most approved shape, and it is important that the pumps have ample power and capacity to fully and continuously feed the jets.

CHAPTER XVIII.

THE WELL SYSTEM OF FOUNDATIONS FOR BRIDGE-PIERS, ABUTMENTS, QUAYS, AND DOCK-WALLS, ETC.

THE well system is particularly adapted for loose sand, mud, and silt, but practice shows it is not suitable for soil harder than ordinary sand, unless under exceptional circumstances. It is an especially good method, if the mud is sufficiently watertight, for the well to be pumped dry, so that the excavation can be executed without dredging machinery. It has many important advantages over a timber pile foundation in sand, silt, or mud, and in warm climates should be preferred. Brick walls have been sunk through clay, boulder sand, and solid beds of shingle. If boulders or *débris* are expected to be encountered in sinking, masonry, concrete, or brick cylinders should not be used, but iron, as they are stronger and more air-tight, as neither dredging nor divers may be able to remove the obstruction, and the compressed-air system may have to be adopted. The well system has failed when boulders and obstruction in sinking have been encountered. The modern practice of well sinking is to diminish the number of wells and to increase the diameter or area of each, as the larger wells are more stable, and the resistance opposed to the lateral force of winds and currents tending to overturn them increases with the diminished number of sub-divisions. The resistance has been stated to be approximately as inversely as the square root of the number of separate parts into which the foundation is divided. It is important that the diameter of a well should bear a sufficiently large proportion to its height. If they are in groups, the wells should be securely tied at the level of the river bed.

Wells have been sunk of 40 ft. diameter, their internal diameter being about 32 to 33 ft. General experience seems to show that, within reasonable limits, the larger the diameter of the well, the less difficulty there is in sinking it, and the height should not exceed about one-fourth of the diameter.

Large brick cylinders are more likely to crack than small wells, but this defect is readily overcome by substantial construction, by having numerous holding-down bolts, and by taking care that the wells sink evenly, and do not hang from surface friction, which resistance may be distributed unequally. Many of the remarks made with respect to the cylinder system and sinking are obviously applicable to the well method of foundations, and therefore are not here repeated.

Trouble has been experienced when two rows of wells have been sunk close together, as they tilt and often become jammed against either the top or bottom of the opposite well; and in the process of removing the material from under the curb, in order to bring it to perpendicularity, the adjacent well tilts, from the earth moving from under its curb at the nearest part. One row of cylinders of large diameter is decidedly preferable to two rows of small wells sunk close together.

Rectangular and oblong wells have been used, and for dock or quay walls it may be necessary to adopt them to produce a straight face, but the circular has been proved by long experience to be the best form, especially in India, where the course of a river may, during floods, be at right angles to the centre line of the usual channel, showing that piers should have equal bases in all directions, and therefore be circular. One of the principal objections to rectangular wells is that in excavating the soil a circular hole is formed, and therefore the wells hang on the four corners, instead of equally all round, and therefore fractures often occur; they are also much more difficult to "right" than cylindrical wells, should they become inclined in sinking. There is less material or steining in a circular than in any other form of well of equal area, and it is also the best for resisting lateral pressure, because a strain at any point in the ring is communicated to the whole.

GENERAL CONSTRUCTION.—As wells may be severely strained during sinking, special attention must be given to their construction. They have to withstand the external pressure of the soil and the head of water, the strains arising during unequal sinking, the tensional strain from surface friction, which may act upon only a small surface of the well, the compressive load from the weight of the well and any kentledge, and rushes of water or soil; for brickwork wells have been burst because of a sudden rise of water inside them; the reason of such a rush of water is frequently that the diggers and dredgers have only excavated a central hole, and because this has extended from 7 to 9 ft. below the cutting edge before the surrounding earth would fall into it. In order to counteract the pressure due to a considerable head of water, it may be prudent, even if the internal excavation can be effected in the open air, to execute it by dredgers or divers, but sinking will then not be so easy, and kentledge may be required. The weight of the superstructure should be equally distributed over the whole area of the top of the well.

The thickness of a well will depend upon the nature of the material used in the steining, the height and diameter of the well, and character of the soil to be sunk through; and sufficient space must be left in the interior for the excavating machinery. It is an advantage if it can be of sufficient thickness so that its weight will cause it to penetrate the earth. Wells of less outside diameter than about 10 ft. are now but seldom used. The thickness of the steining usually varies from 2 ft. 3 in. to about 4 ft., but, for quay walls, masonry in Portland cement mortar wells, 33 ft. by 22 ft., 7 ft. 3 in. in thickness, have been sunk at Havre, and are exposed to the sea. When brickwork is used, it is sometimes made to gradually increase in thickness by half-brick projections on the inside to its maximum at the curb seat, but the corbelling is objectionable if dredger-machinery has to be employed for excavating the interior earth, as the dredgers may be caught by the projections. The brickwork is often tapered towards the shoe, so as to offer the least obstruction to penetration. Ordinary bricks, not radiated, are sometimes used for the steining; but radiated bricks are preferable, and they require less cement in the joints. It has been found that bricks exceeding 9-in. ordinary bricks are too large to be economical for the steining. The bricks are sometimes made so that the angle of divergence at the ends, and the radius of curvature of the sides are of the mean radius of the steining. Vitrified face bricks are occasionally introduced if ordinary bricks are likely to be injuriously affected by the salt water in the soil. The brickwork should be thoroughly bonded together, and its outer cylindrical surface smoothly plastered with Portland cement, so as to lessen the surface friction. It is sometimes bonded with hoop-iron laths, at intervals of 3 or 4 ft. On adding fresh brickwork to that already made, in order to get a good joint, care should be taken that a clean surface is obtained to which the cement mortar can firmly adhere.

The steining should be allowed to stand until the masonry, concrete, or brickwork is thoroughly set. Cases have occurred in which fracture and failure of a well have arisen because the material had not had time to fully set. Concrete made of about equal portions of cement and other constituents should have at least six or seven days to consolidate and set; and masonry, brickwork, and Portland cement concrete in the proportion of 5 to 1, not less than fifteen days, and one month is to be preferred. The cement should always have a high cementitious strength. Good Portland cement concrete steining appears to be a preferable material to brickwork or masonry, on account of its homogeneity, and it is also cheaper and heavier than brickwork. The thickness of the concrete is usually from one-fourth to one-fifth of the outside diameter, a minimum thickness being 2 ft. to 2 ft. 6 in.

An objection to brickwork, unless the bricks are of a non-porous nature, is that when such a well is empty, water percolates from the surrounding earth through the steining if the depth sunk is considerable. It is therefore advisable to test any bricks before using them in the steining in order to ascertain the head of water they can resist. An outside thick rendering of Portland cement may prevent any percolation. Masonry wells have been sunk from 8 to 10 ft. into the ground in shallow depths of water, such as from 10 to 15 ft. The masonry being about one-fourth of the diameter in thickness or 3 ft. 3 in. for a 13 ft. outside diameter well, the lower portions being set in cement, the curb being of wood, 4 to 6 in. in thickness, strengthened by angle-irons bolted to the steining.

Tie-bolts are an essential in the steining. The probable strain on these bolts would be, during the forcing-down operations, if the wells hung, that of the surface friction less the weight of the well, the kentledge, and the tensional resistance of the steining in some measure. It usually happens, when a cylinder is suspended by surface friction, that the resistance is almost entirely on the ring of the well just added. The vertical tie-bolts are mostly from 1 to $1\frac{1}{2}$ in. in diameter, and are placed at intervals of from 3 ft. 6 in. to 6 ft. round the whole central circumferential line of the steining. They must go through the steining from the curb seat to the top of the well, and be securely attached to the curb. Rings of flat iron, 4 to 6 in. in width, and about $\frac{3}{8}$ in. in thickness, through which the tie-bolts pass, are sometimes introduced at intervals of 8 to 10 ft., in the height of the well. This plate is also occasionally of a width equal to the thickness of the steining, and can be cottered down to the tie-bolts; or the latter can be in lengths, with a nut or coupling 6 or 8 in. in length, with a large washer, the nut or coupling being screwed down tight on the completion of each length of the well, additional lengths being screwed on as fresh rings are added. The latter method is to be preferred to any cotter arrangement, and there is an advantage in this system of tie-rod, nut, and washer, because each length of well is compact in itself, and is also joined to other rings. If from any reason, such as grooving the joints of the different lengths, it should be inconvenient to place the tie-bolts in the centre of the steining, they should be fixed nearer the outer circumference than the inner surface, as the greatest resistance is encountered on the outer circumference owing to surface-friction. On the upper surface of the highest length of the well, a plate about ⅜ in. in thickness should be placed, upon which to screw up the tie-bolts tightly; the well is then complete. In order to prevent any fracture caused by iron bolts, bars, etc., being built in the steining, it may be necessary to provide for alterations in their length, consequent upon variations in the temperature.

The remarks made on the best form of cylinders are applicable to wells, but there are even stronger reasons for the adoption of the circular form for wells. Oblong and elliptical wells have been sunk, and a straight face for a wall may be necessary; if not, the general testimony is decidedly in favour of the cylindrical form. Circular are far preferable to oblong wells, and elliptical are better than the latter, although that form is much inferior to the cylindrical. Among the objections to oblong, and in a lesser degree to elliptical wells, are the great practical difficulties of obtaining a thorough bond at the corners; the long, straight side walls which have to resist the pressure of the soil, the great trouble experienced in sinking them evenly, and righting them should they tilt as frequently happens. This form should therefore not be adopted, unless a straight face must be afforded; but of course, if the depth to which the wells have to be sunk is small, as is usually the case in quay or dock walls as compared with bridge-piers, these objections are not nearly so important. If it is absolutely necessary to approach such a figure, the elliptical should be used; but, if possible, the circular. There is seldom any necessity for the employment of any other form than the circular for bridge piers, as several wells can be sunk in a line, which can be built upon in the usual shape of a pier, but the wells must not be too close together. When large wells are sunk, with only about 3 ft. clearance, small arches are sometimes turned over them, so as to make a continuous surface, upon which the pier is built, or the wells are corbelled out at the top, and are connected by two or three thick courses of ashlar being placed upon them. The difficulty of sinking is much increased by grooving the wells or interlocking them, as the slightest divergence from perpendicularity retards the sinking.

CURBS.—The curb should not present a large flat surface in order that the resistance to descent may not be great. In the previous notes on the cylinder system a union of the metallic and non-metallic methods is referred to; a modification of this combination might be effected by a lengthening of the outer edge of the curb, which would then not only give a more pointed cutting edge, but would to some extent prevent the exterior soil being forced into the well. Cast-iron are heavier than wrought-iron curbs, but in material which admits of easy penetration, such as sand and mud, a heavy curb is a disadvantage, because in addition to the expense of construction, it brings more weight on the foundation with no corresponding gain. Curbs require to be sharper and longer, according to the compactness and hardness of the soil to be penetrated. The form of the curb is usually an inverted right-angled triangle for soft soils, the perpendicular being about $1\frac{1}{4}$ to $1\frac{1}{2}$ time the width of the base, upon which latter the steining is built. If the steining is Portland cement concrete, the thickness is sometimes reduced by one-half at the curb seat, it being bevelled for about 2 ft. upwards from it on the outside, thus lessening the width of the curb.

The curb should be strengthened and stiffened by gusset-plates and angle and T-irons, and should not be very thin, so that it may be heavily loaded ; and should it meet with boulders or obstructions, that it may be of sufficient strength to resist them, as the expense of a little extra iron is nothing as compared with the cost entailed by fracture of a well during sinking operations. Iron is the best material for a curb, but it may not always be obtainable, and time may be saved by using some other substance ; if such should be the case, either wood or concrete can be adopted, both of which should, however, be shod with iron. The triangular open space formed in an iron curb is generally filled with concrete. In designing the curb it should be borne in mind that if it bends it will most probably be destroyed, or cause an entire suspension of operations for a time. Its dimensions and thickness will vary according to the thickness of the steining, height and diameter of the well, the hardness of the soil, and the probability of obstruction being encountered during sinking. In wells of moderate diameter, the pressure per square foot on the curb, assuming it acts equally over the whole surface, seldom exceeds from about 3 to 5 tons. To bind the steining to the curb, the cutting-plate should be carried up above the level of the horizontal plate, from 3 to 6 in., which will prevent the steining slipping off on the outside; an angle-iron riveted to the horizontal plate on its inside diameter will hold it internally. Such a curb of ordinary diameter weighs about 3 to $4\frac{1}{2}$ tons. Cast iron curbs are generally constructed in segments bolted together.

If the curb is made of timber, the wood should be hard, and have a comparatively high resistance to crushing, such as oak, beech, ash, American plane, sycamore. The usual method is to build up wedge-shaped segments, which should break joint and be fastened together with strong bolts. The end of the wedge, which can be about 3 in. in width, should be protected by plates, and a cutting-plate should project for about 1 ft. below the wedge-end, upon which a strong angle-iron should be bolted, and the cutting-plate be riveted to it so as to make it as rigid as possible. The area of the wedge-end must be regulated according to the strength of the timber used, and the weight that it will have to bear, not only from the well, but also from the kentledge.

SINKING NON-METALLIC CYLINDERS.—When wells have to be sunk in a waterway, the simplest mode of pitching the curbs is to form an artificial island. It may not be always necessary to make such islands if the water is of less depth than 5 or 6 ft.; and, with plenty of tackle, the curbs and first lengths of the well might be pitched in an ordinary current where the depth of the stream does not exceed 10 ft., but much will depend upon the diameter of the well; if of small size, it would very probably upset should the diameter be less than the depth of the water; and rather than risk this, it may be advisable to raise a bank by simply casting earth into a river until it reaches above the water level. There are several methods of making these artificial islands. In still water, circles of sandbags can be deposited by divers, or other means around the site of the wells forming the pier, and when the circle is completed, sand can be deposited until a level of about 1 ft. above the water is reached. If the river has a gentle current, sandbags can be laid on the down stream end of the site of the pier, and on the two sides of the site, forming a three-sided wall; sand must then be deposited at or about the centre of this enclosure, and the current will throw it against the down stream sandbags, at which end a bank will be gradually raised to the surface of the water; when this occurs, it is usual to deposit the material on the upstream open end of the enclosure, whence the current transfers it to that part of the bank previously formed, and gradually the artificial island is completed ready for the wells to be pitched. In a swift current, this system would scarcely afford sufficient stability. Piles must then be used as a protection and auxiliary to the sandbags. Spurs made of trees and stones, etc., are sometimes placed in swift rivers to divert or slacken the current during the construction of an artificial island. There are obviously many different ways of making an artificial island of sandbags and piles. A successful method is to drive piles on the upstream side of the river, which are also used as staging; these piles are driven so as to make a cut-water; gunny bags are then placed by divers on the remaining sides until the whole area of the pier is enclosed. On the down-stream side there is no danger of the bags being washed through, but the sides parallel with the stream should have piles driven about 2 to 3 ft. apart for the bags to rest against. The island is then completed by bags being placed by divers, the interstices between the bags being filled with loose sand until the water level is reached.

At the river bridge over the Ems at Weener, wells which were 13 ft. 1 in. diameter for a height of 8 ft. 3 in. from the curb were built inside a wooden sheathing, and the bolts suspending them from the stage during construction and lowering were built into the brickwork. They were circular at the bottom and gradually tapered inside towards the top, which is at low-water mark. The depth of water was only 7 ft. 6 in. The thickness of the well on the curb seat was 2 ft. 1 in., at the top 1 ft. 3 in. At low-water mark the ordinary brick well begins. They were built in cement brickwork, one of cement to two of sand, faced in river piers above bed of river with alternate courses of whole and half bricks in cement, one of cement to one of sand. The piers stand on the filled up wells and are faced with hard brickwork. The wells were sunk from a floating stage carried by two barges, between which there was just room for the cylinder. They were hung by bolts and links from a circular timber frame and were sunk on the ebb tide. This example is mentioned, as it is an alternative method to adopt to that of erecting wells on an artificially formed island. Circumstances must decide which plan of operations is to be preferred.

Wells for foundations are often sunk in the silty or sandy bed of rivers, which become dry in summer, and where there is therefore no running water to contend with. The usual method of sinking wells is first to excavate the ground to the water level, and then to lay a curb on the soil. To keep the curb in place it is advisable to sink it to the level of its top plate, or seat, before commencing building operations, after which the tie-rods can be fixed and the first length built. In order to obtain a perfectly vertical descent, and to enable the direction of the sinking of the well to be easily corrected, it is prudent to build the first or curb length of much less height than the remaining lengths ; the second length can be made of greater height, and the third and other rings of a convenient length. The first or curb length can therefore be about 5 ft. in height, the second 8 to 10 ft., and not exceeding the latter dimension, and the third not higher than 15 ft. These heights are for wells from about 12 ft. in diameter. It is generally agreed among engineers experienced in well sinking that it is of paramount importance the curb and the first two lengths should be perfectly vertical ; the after sinking is then a comparatively easy matter with ordinary care. To ensure the curb being always vertical, the steining should be equally built around its whole circumference ; and in sinking the curb and the first and subsequent lengths, the material from the interior should be methodically excavated either evenly over the whole internal area, or in the centre, the former being the better system. There are objections to increasing the height of the wells in order to obtain greater dead weight for sinking purposes, which become more cogent in the case of wells of small diameter. Among them may be named the additional height of the staging for building the lengths, the increased range of the tackle required in lifting the material for the rings, the greater difficulty of righting the well should it tilt, the augmented height through which the excavating machinery has to act, and the less stability. At the Plantation Quay on the Clyde, Messrs. Bateman & Deas instead of first depositing the curbs and then building upon them when *in situ*, constructed the concrete rings in frames on a platform near the line of quay, and had them put together on the curb after they were consolidated. By the adoption of this latter method of adding lengths much time is saved, and the rings can be attached almost as quickly as iron rings, the joints being simply cemented together. If proper joints are made between the lengths, it appears to be the better system to adopt, especially after the first length.

The number of wells commenced in one season should be such that they can be completed before the time of deep flow or floods, so as to avoid running the chance of their overturning, but should any be incomplete, the tops should be taken off at low-water level, and the site indicated.

In well sinking the use of compressed air is exceptional, the soil being removed by dredging. The pneumatic system of sinking is generally only used when all other methods have failed to remove unexpected obstructions met with in sinking, and no other course is open. Every endeavour should be made to stop the leakage of the air by the use of impermeable coatings. The concrete, or, if the well is of brickwork or masonry, the cement mortar should be allowed ample time to set and harden, and the joints and steining should be thoroughly grouted with liquid cement. Records do not show many successful applications of the pneumatic system to sinking non-metallic cylinders under a pressure of more than about 40 ft. head of water. At Rochefort, the rubble-masonry-set-in-cement-mortar-wells were so constructed that the compressed-air system could be used when there was a rush of silt; a recess was made round the inside of the well at a height of about 16 ft. from the bottom; this was used as the springing of a vaulted roof of masonry in cement 3 ft. 3 in. in thickness. Concrete was deposited upon the top of the arch, a circular 2 ft. 4 in. opening being left in the centre. The space below the roof provides a working chamber to which the hollow cylinder gives access, an air-lock being fixed at the top of the wall over the cylinder. The soil was soft alluvium.

Until a depth of water of about 5 or 6 ft. is reached, the soil can be excavated by men with scoops or a jham, then, if necessary, by dredger-excavating apparatus. (See Chapters XV., XVI., and XVII.)

The notes on cylinder sinking are generally applicable to well sinking. Should firm strata be encountered in sinking, the well must be weighted in order to make it descend. Kentledge placed upon the well is to be preferred to building the well to an additional height. When water percolates through the steining of a well it has the effect of drying the earth around it, the consequence being that surface friction is increased and the well harder to sink, and in some soils the result may be that around the outer circumference of the well the earth may become dried, detached from the wetter earth, and adhere to the well in its descent. On meeting with obstructions in sinking, the water, if possible, should be removed from the well; this may perhaps be effected by bags or pails, if not, by pumping. As stated under the head of cylinder sinking, clearing the well of water will often make the sand move at the bottom because of the unbalanced exterior pressure, and "blows" may occur which may be attended with danger to the well. Perhaps the quickest plan to adopt is to send down helmet divers with picks and jumpers when satisfactory progress is not made in the sinking.

It will usually occur that the quantity of material excavated from a well is from about 60 to 100 per cent. more than the cubic contents of the subterranean portion of the well, although this excess may be much more in very unstable earth; but if wells sink freely, and the water-levels inside and outside are about equal, the extra excavation may not exceed 30 to 40 per cent. It is advisable to proceed with the sinking continuously by day and night shifts in rivers where the working season is short.

From an analysis of many works, the rate of sinking appears to decrease according as the depth increases. The rate of sinking in sand with ordinary obstacles, of some wells 13 ft. 6 in. in outside diameter, was from 12 to 18 in. per day at starting, up to about 10 ft. deep, and 6 in. per day for a depth of 20 ft. Of course the rate will vary considerably according to many circumstances, which have been mentioned in cylinder sinking, and the progression is frequently very variable. Ten to twelve men and a foreman will be sufficient to effect this amount of sinking for a well 10 to 12 ft. in diameter, or one man for every foot of diameter of the well. After a depth of about 30 ft. has been reached, unless the soil is favourable, the amount of sinking in a day of twenty-four hours will often become tedious, perhaps not exceeding 2 in. and even 1 in. Wells 8 to 13 ft. 6 in. in diameter have been sunk at the rate of from 3 to 6 in. per hour. At the Sone Canal, where most of four thousand wells were sunk from 8 to 10 ft. in the sand, a well 10 ft. by 6 ft. was sunk 10 feet in nine hours; but this is an exceptional rate, and in any soil other than most open sand, it is hardly ever reached. Wells sometimes go down as much in one hour as shortly after will take a day, therefore they require to be thoroughly watched; but progress of course much depends on the efficiency of the excavating apparatus employed.

From an examination of several examples where seams of clay have been encountered after sinking in sand, and a comparison of rates of the same diameter of well and depths in clay and in sandy soil under analogous circumstances, it appears that the rate of sinking in the former is from about one-fourth to one-fifth of the latter.

In sinking groups of wells joined together, it has been found that the excavation in all the wells of one cluster should be carried on simultaneously and equally, and then there is no difficulty in sinking them evenly, and that less kentledge is required to force them down.

To "right" or bring back to perpendicularity, a well which has tilted after the second length has been sunk, is an operation necessitating great care, hence the importance of the descent always being kept vertical. Wells have been righted when the third length has been built, by pulling them over during sinking operations. After they have penetrated more than 20 ft. it is very difficult to right them, and after wells of ordinary diameter have sunk to a depth equal to twice their diameter, it is almost impossible. Among the expedients used and suggested for bringing wells back to perpendicularity are, the insertion of perforated pipes from which water is discharged under pressure on the higher side to diminish the surface-friction, additional weights on that side of the well, which must be done very carefully so as not to unduly strain the steining, excavating by a dredger outside the elevated portion, by having a pulling-strain in the direction in which the well is to be righted, by excavating under the curb, by strutting on the lower side, and by depositing stiff material on the surface of the ground outside and close to the well. (See also Chapter VII.)

After a well has been sunk, in depositing the hearting a thorough connection must be effected between it and the steining, so that the whole may act as a monolithic mass. This can be attained by toothing the internal face of the steining, if of brick, or by set-offs if of brickwork or concrete, and in the interior diameter of the well the concrete face should be rough, so as to bind and bond with the hearting, but, of course, the external face must be as smooth and as even as possible, to lessen surface-friction during sinking. The concrete should not be simply thrown in from the top of the well, if the latter is dry, but be gently deposited from a moderate height, so that in the descent, the heavier may not separate from the lighter constituents, and it is best when merely shovelled from a stage practically on the level with the surface; or it should be lowered in self-acting discharging skips if it has to be passed through water. (See Chapter IX.) When the wells are finished, the spans should be measured to ascertain what provision must be made for lateral and longitudinal divergence.

For quay and dock walls not subject to very heavy loads, it is not necessary that the whole of the hearting of the wells should be of concrete, as, provided the bottom of a well is sealed from water, which a layer of from 5 to 12 ft. of cement concrete generally effects, and the concrete bottom has a proper bearing, sand will do; but it should be damp when deposited, and well rammed, and water should be prevented from accumulating behind the wall of wells, as then the pressure may become

THE WELL SYSTEM OF FOUNDATIONS.

that of the head of water. For practical information on concrete, see the second edition of the author's book "Notes on Concrete and Works in Concrete."

At Pallanza, Lake Maggiore, brickwork wells 23 ft. apart, centre to centre, 7 ft. 6 in. external, and 5 ft. 3 in. internal diameter, were used. They were sunk into hard compact sand, 20 to 30 ft. below low water, and filled with concrete. Upon them nearly semi-circular arches were turned, and the quay wall built. It was stated that whereas all the other walls on the shore in the neighbourhood were more or less damaged, and some collapsed, this system has proved entirely satisfactory. Quay walls on this principle have among other places been erected at Bordeaux, St. Nazaire, and Rochefort, where semi-circular arches, 30 ft. span, were turned on piers 16 to 20 ft. in thickness, sunk to a depth of 50 to 90 ft. in soft alluvium. The wells were of rubble masonry in cement mortar, and were built in a trench to a height of 10 ft., and allowed fifteen days to set, when the excavation was commenced by which means they were sunk.

Bridge abutments have, especially in India, also been built on arches turned upon well-piers a few feet below the ground.

INDEX.

A.

Abutments, Well System of, 129.
Air, Compressed, Effect on Men, 81-83.
— — Cooling, 78, 79, 81.
— Noxious, in Cylinder Sinking, 49.
— Penetrating Pressure on some Substances, 74, 75.
— Quantity required, 73, 75, 76, 78.
— Temperature, Precautions, 80, 81.
Air-Compressors, Arrangement and Character, 77-79, 81.
— Hydraulic, 79.
— Old, 78.
— Number of, 77.
Air Escape, Effect of, 74.
Air-Lift, Excavating Apparatus, 116.
Air-Lock, Arrangement and Purpose, 83-87.
— Doors, Interlocking, etc., 84-86.
— Floor, 77, 84.
— Height, 84.
— Light, 83.
— Lighting, 84.
— Position of, 86, 87.
— Removal of Excavation in, 86.
— Roof, 77.
— Signalling Apparatus, 81, 85, 86.
— Size of, 84, 85.
— Temperature, 80-81.
Air-Pipes, 84.
Air-Pressure, Advisable, 73-76.
— Declining, Effects of, 76.
— Formula, 75.
— when Injurious to Men, 80, 81.
— when Springs Encountered, 76.
— Working-hours in, 80.
Air-Pumps, Working Continuously, 73.

Air-Supply, Constant Discharge, 74.
— Purifying it, Necessity of, 74, 81.
Air-Tightness of a Cylinder, 73.
Air-Waste in Sinking Cylinders, 73, 74.
Areas of Cylinder, Table of, 18, 19.
Artificial Islands, Pitching Wells upon, 124.

B.

Bag and Spoon Dredger, when Useful, 101, 104.
Base of Cylinder Piers, Load on, 24-31.
"Blows" of Soil during Sinking, 46-48, 54, 71, 101.
Bracing Cylinder-Piers, 3, 4.
Brickwork, Penetrating Air-Pressure, 75.
Bruce and Batho's Excavators, 106.
— Special Features, 101, 106.
Bucket-Dredgers, Construction, 100.
— Defects, 96, 97, 99, 101.
— Discharge, 100.
— Form, 100.
— v. Grab-Dredgers, 99.
— Hoisting Apparatus, 102.
— Improvements in, 101, 106, 110.
— when Ineffectual, 94, 97, 106, 108, 109.
— v. Jumpers, 96.
— Lifting Power required, 98.
— Shape of, causes Success or Failure, 99-101.
— Size of, 97, 98, 100.
— Suitability of Earth for, 95, 96, 99, 101, 109, 110.
— Working Capacity, 99, 101.

Bull's Dredger, 108, 110.
Bursting of an Air-Lock, 83.

C.

Compressed-Air System of Sinking, 70-72.
— Adoption, 94.
— Advantages, 72.
— Dangerousness, 73.
— Depth of Economic Adoption, 71-73.
— Disadvantages, 72.
— and Dredging, 94, 95.
— Effects of, 70, 71, 81, 83.
— Limiting Depth, 73.
— Old and New, 70.
— Precautions, 81.
— Requirements, 70, 71.
— Signalling Apparatus, 81.
— Waste of Air in, 73.
Compressors, Air, Arrangement and Character, 77-79.
— Number, 77.
Cooling Compressed Air, 78, 79, 81.
Cutting-Edge, Excavating around and under, 96, 97, 102, 104, 106.
Cutting-Ring, 5, 6.
Cylinder-Piers, Advantageous Application, 1, 2, 4.
— Area required for Excavating, 63, 64, 97, 98.
— Compressed-Air, Method of Sinking, 70-72.
— Cost of Different Methods of Sinking, 71.
— and Cribwork, to Resist Ice-Floes, 2.
— Diameter required, 10-17.
— Distance between, 5, 44, 45.
— Forces Governing Stability, 23, 24.
— Increased Stability, 4, 5.
— Large v. Small, 5, 8, 9.
— Limiting Height, 13, 14.
— Load on Base, 24-33.
— and Opening-Bridges, 4.
— Position, 5, 8.
— Shortening Spans, 5.
— Sinking, 40-50.

Cylinder - Piers, Surface Friction, 33-39.
— Unsuitableness, 3.
— Water-tightness, 67.
— Weights, Diagrams and Formulæ, 11, 12.
— — Table of Areas, etc., 17-23.
— v. Wells, 1, 2.

D.

Design of Bridge-Pier, Determining, 1, 2, 24, 25.
Diack's Excavator, 108.
Diameter required, Cylinder Bridge-Pier, 60-62.
Dimensions of Rings, 6, 7.
Discharging Material through Air-Lock, 85, 86.
Distance between Cylinder-Piers, 5.
Diver's Helmet, Muddy Water, 82.
— Precautions, 82.
— Rough Weather, 82.
— Signalling, 81.
Dock Walls, Well System of, 116, 126, 129.
Doors, Interlocking Air-Lock, 85, 86.
Dredging Apparatus for Cylinders and Wells (see also Excavating Apparatus).
— Advantages, 91.
— Bag and Spoon, Useful Application, 101, 104.
— Bucket, when Ineffectual, 94-97, 106, 108-110.
— v. Compressed-Air System, 94, 95, 110.
— Construction, 100, 101.
— Defects, 96-98.
— Design and Purpose, 90 - 93, 95.
— Disadvantages, 91, 92.
— Expulsion System, 117.
— Grab-Dredgers, 99.
— Hoisting and Discharging Apparatus, 102, 103.
— Improvements in, 100, 101, 106, 110.
— Jumper and Bucket System combined, 93-95, 99, 105, 109.

INDEX.

Dredging Apparatus, Lifting Power required, 98.
— Shape of, causes Success or Failure, 99-101.
— Single v. Double Chain, 110.
— Size of, 97-100.
— Suitable Earth for, 95, 96, 99, 101, 109, 110.
— Water-jet, 116-118.
— Working Capacity, 99, 101.

E.

Excavating, Working Area for, 63, 64.
— by Compressed Air, 116.
— Effective Action, 116.
— Expulsion System, 117.
— Water-pressure System, 117, 118.
— — when Ineffectual, 117.
Excavating Apparatus for Cylinders and Wells (*see also* Dredging Apparatus).
— Adoption, 94, 95, 110.
— Advantages, 91, 92.
— Defects, 91, 92, 96, 97.
— Design and Purpose, 90-93, 95.
— Hoisting and Discharging Apparatus, 102, 103.
— Improvements in, 101, 102-106, 110.
— Jumper and Bag Dredgers, 108-109.
— — and Bucket System combined, 93-95, 99, 105, 109.
— Lifting Power required, 98.
— Screw for Loosening Soil, 107.
— Shape of, causes Failure or Success, 99-101.
— Size of, 97-100.
— Suitability of Earth for, 95, 99, 101.
— Various, Notes on, 103-111.
— Working Capacity, 101.
Excavation, Discharging Pipe through Air-Lock, 85, 86.
— Method of, in Cylinder, 87.
— Removal of, in Air-Lock, 86.
Excavator-Pumps, 111, 112.
Explosives for Removal of Obstructions, 55.

F.

Floor of Air-Lock, Pressure on, 77.
Flotation Power of Cylinder, 19-23.
Form of Cylinder Rings, 5, 8, 9.
Foul Air in Cylinders, 49.
Fouracre's Dredger and Spider Clay Cutter, 106.
Frictional Resistance, Earth on Cylinders, 88-89.

G.

Gatwell's Excavator, Special Features, 105, 106.
Grab-Dredgers v. Bucket-Dredgers, 99-100.
— Construction, 100, 109.
— Object, 99, 109.
Gradual Pressure Room in Air-locks, 83.

H.

Hearting, Contraction, etc., 67.
— Depositing in Compressed Air, 67.
— — under Hydrostatic Pressure, 67.
— Deposition, 66, 69.
— Grout, for filling Cavities round the Cutting Edge, 66.
— Purpose of, and Material for, 65, 66.
— Surface of Base before Deposition, 66, 111.
— Watertight, Importance of, 67, 68.
Height, Limiting, of Cylinder Pier, 13, 14.
Hoe-Scoop, or Indian Jham, Usefulness of, 104.
Hoisting and Discharging Apparatus for Dredging Machinery, 102-103.
Hydraulic Method of Removing Obstructions, 56.

I.

Inclined Cylinders, A Cause of, 57.
— "Righting," 56, 57.

Iron, Cast, Penetrating Air Pressure, 74.
Iron Cylinders, Advantages of, 1, 2.
— v. Well System, 1, 2.
Iron Rings, Dimensions of, 3, 4.
— Object of, 3.
Islands, Artificial, for Pitching Wells upon.
Ive's Excavator, 107.

J.

Jham, Indian, or Hoe-Scoop, Usefulness, 104.
Joints, Cylinder Rings, 6, 7.
Jumper, Chiselled Rail, 105.
— Rail, 108.

K.

Kentledge, Calculation, 60-65.
— Cast, 58.
— Large and Small Cylinders, 59, 60.
— Methods and Precautions, 58, 59.
— Permanent Hearting for, 59, 60.
— Quantity required, 64, 65.
— Resistance to be Overcome, 61, 62.
— Thickness and Weight of Casing required, 62, 63.
— Top, Inside, and Outside, 58, 59.
— Unobstructed Area required in Cylinder, 63, 64.
— Water-tank, 58, 59.

L.

Large Cylinders v. Small, 5.
Leakage of Air in a Cylinder, 73, 74.
Leslie's, Sir Bradford, Rotary Plough and Boring-head, 111-113.
Lifting Apparatus, Dredging Machinery, 102, 103.
Lighting, Air-Lock, 84.
— Working Chamber, 88-90.
Limiting Depth, Compressed-Air System, 73.
— Height of Cylinder, 13.

Load on the Base of a Cylinder Foundation, 24-28.
— Safe, 26-31.
— Table of Safe Loads, 31-33.

M.

Machinery, Air-Compressing, 77.
— Character and Arrangement, 78.
Making-up Ring, 7.
Milroy's Excavator, 107.
Molesworth's Dredger, 111.
Mooring Pontoons, 51, 52.

O.

Obstructions in Sinking,
— Advantage of Compressed-air System, 54.
— "Blows" of Soil caused by, 101.
— Explosives, Use of, Questionable, 55, 56.
— Hydraulic Method of Removal, 56.
— Methods of Removal, 53-57.
— Removal of, 54, 95.

P.

Pipes, Air, 84.
Pontoon System of Floating-out Cylinders, 51, 52.
— Improvement in, 52.
— Precautions necessary, 51-53.
— Small Cylinders, 52.
— Top and Bottom Guidance necessary, 51.
Portland-cement Concrete Cylinders, 3.
Position of Air-lock, 86, 87.
— Cylinder Piers, 5.
Pressure, Normal, of Soil, 28.
Pumps for Sandy Water, 111, 112.
Purification of Compressed Air, 81.

Q.

Quay Walls, Well System of, 116, 126 129.

R.

Rammer for Hard Soils, 104. 105.
Reducing Ring of Cylinder, 4.
" Righting " an Inclined Cylinder, 56, 57.
— an Inclined Well, 125, 128.
Rings of Cylinder, Contraction of, 67.
— Cutting-Ring, 7.
— Dimensions of, 6, 7.
— Form of, 5.
— Making-up Ring, 7.
— Thickness of, 6.
Roof, Air-Lock, Pressure upon, 77.

S.

Sand-Pumps, Arrangement and Use of, 111, 112, 114.
— Combined with Cutters, 111, 112.
— Ead's, 116.
— Kennard's, 108.
— Working, 111, 112.
Sandstone, Penetrating Air-Pressure, 75.
Shafts to Working Chamber, 85, 87, 88.
Signalling Apparatus, Air-lock, 85.
— Compressed-Air System, 81.
Sinking Cylinders, "Blows" of Soil during, 46, 49.
— Close together, 44.
— Comparing Cost of, 40.
— Compressed-Air Method, 70, 72.
— Cost of Different Methods, 40, 71, 73.
— Danger of Sudden Sinking, 44.
— Earthbound, 48, 49.
— General Notes, 40.
— Loose Soils, 48, 49.
— Methods of, 40-42.
— Noxious Gases Encountered in, 49.
— Obstructions in, 53.
— Outside Subsidence caused by, 46, 48.
— Pontoon, System of Floating-out, 51, 52.
— Precautions in, 45, 46.
— Staging, 50, 51, 102.
— Suddenly to be Avoided, 71.
— Vertical Sinking Important, 43, 44.
— Water-levels, Inside and Outside, 47, 48.

Sinking Wells by Compressed Air, 126.
— Discontinuing, When Incomplete, 126.
— General Notes, 124-128.
— Rate of Sinking, 127.
— " Righting " Them, 125, 128.
Small v. Large Cylinders, 5, 8, 9, 44.
Spider Clay Cutter and Dredger, Fouracre's, 106.
Stability of Cylinder-Bridge Piers, Forces Governing, 23, 24.
Staging, 9, 50, 51, 97.
— in Exceptional Circumstances, 50, 51.
— Fixed, when Impracticable, 50.
— Use of, 50.
Stoney's Helical Excavator, 111.
Strong's Excavator, where Useful, 105.
Subsidence of Ground during Sinking Cylinders, 46.
Supply Shafts, 87, 88.
Surface Areas of Cylinders, Table of, 19-22.
Surface Friction, on a Cylinder Pier, 33-39.
— Depth, Influence, 37, 38.
— — Reliable, 38.
— Dry and Wet Surface, 37.
— Large v. Small Cylinders, 38.
— Table of Values, 39.

T.

Telescopic Dredger, Furness and Slater's, 111.
Temperature of Air, Precautions, 80, 81.
Thickness of Rings, 5, 6.
Timber, Penetrating Air-Pressure, 74.
Top Ring of Cylinder, 7.

W.

Water-jet Excavators, 112-118.
Water-sealing a Cylinder, 68, 69.
— Thickness required, 69.
Weight of Cylinder Piers.
— Diagram and Formulæ, 10-12.
— Tables of, 17-23.

Well System, Abutments, 129.
— Advantageous Application, 1, 4, 118.
— Conditions when it may Fail, 1, 118.
— Curbs, 122.
— — Heavy v. Light, 122.
— — Material for, 122, 123.
— — Securing Steining to, 122.
— Excavation, Internal, 127.
— Form of Wells, 119, 122.
— Grooving and Interlocking Wells, Objections to, 122.
— Hearting, 128.
— Height of Rings, 125.
— Large v. Small Wells, 118, 119.
— Methods of Sinking, 124, 125.
— Modern Practice, 118.

Well System, Quays and Dock Walls, 116, 126, 129.
— Sinking Wells, 124.
— — Close together, 119, 128.
— Steining, 120-122.
— Strains on Wells, 119.
— Tie Bolts, 121, 122.
Wood, Penetrating Air-Pressure, 74.
Working Chamber, Arrangement and Construction, 87.
— Height and Size, 87.
— Lighting, 88-90.
— Purpose of, 87.
— Supply Shafts to, 87, 88.
— Temperature, 80, 81.
— Water Ingression, 88.
Working-hours under Air-Pressure, 80.

Crown 8vo, cloth, 2s. 6d.

SCAMPING TRICKS
AND
ODD KNOWLEDGE

OCCASIONALLY PRACTISED UPON PUBLIC WORKS.

CHRONICLED FROM THE CONFESSIONS OF SOME OLD PRACTITIONERS.

By JOHN NEWMAN,
Assoc.M.Inst.C.E.

REVIEWS OF THE PRESS.

ENGINEERING NEWS (New York).

"This readable and interesting book is arranged as a conversation between two old sub-contractors, in the course of which they deliver themselves of numerous yarns relating to methods practised on various kinds of works, to deceive the engineers and obtain the much-desired 'extras,' thus indicating some of the points to be especially looked after in superintending the construction of works. A still more interesting and valuable feature of the book, however, is that it is full of practical hints and notes upon different methods of carrying out different kinds of work under varying circumstances, giving also advice as to the merits of the different methods."

THE BRITISH ARCHITECT.

"We take the following story from a series of amusing narratives of 'Scamping Tricks and Odd Knowledge occasionally practised upon Public Works.'"

INDUSTRIES.

"This book is out of the run of ordinary professional works, inasmuch as it is intended, not so much for the purpose of showing how public works are to be carried out, as to point out some of the tricks which are practised by those who do not wish to carry them out properly, and to name some methods, founded on practical experience, adopted by sub-contractors and others to cheaply and quickly execute work.

"The young engineer or inspector will find many things in the book which will at least cause him to pay attention to special points in the different departments of civil engineering construction. Such matters as piles, which are chiefly hidden from view, seem to require careful inspection, and in fact all work which is covered up when the structure is completed."

INDIAN ENGINEERING.

"This is an entertaining little book. It abounds with stories of gross cheating. Its publication is not likely to corrupt the morals of native contractors, some of whom could give points to Bill Dark (who is recounting his 'dodges'), inasmuch as that worthy claimed to own a conscience, though it is not very prominent, and always to draw the line somewhere—always put some lime in his mortar, and some headers in his masonry.

"The ingenuity displayed in hiding the results of some of the frauds may be useful in setting young engineers on their guard against the over-plausible."

THE ENGINEER AND IRON TRADES ADVERTISER (Scotland).

"The somewhat uncommon title of this book will in itself prove a ready attraction to the ordinary student of current literature. The title page alone is characterised by a curious vein of humour. The author has, however, a serious and a most important object in view.

"There is a peculiar charm in it not usually found in works where technical details require to be recorded. The many 'dodges' indulged in by these ideal contractors will come as 'eye-openers' to those unacquainted with the subject. We have no hesitation in saying that the volume before us is likely to serve a good purpose, and it is deserving of a wide circulation."

E. & F. N. SPON, 125, STRAND, LONDON.

Crown 8vo, cloth, 7s. 6d.

EARTHWORK SLIPS AND SUBSIDENCES UPON PUBLIC WORKS.

By JOHN NEWMAN, Assoc.M.Inst.C.E.

REVIEWS OF THE PRESS.

ENGINEERING NEWS (New York).

"The book is of a practical character, giving the reasons for slips in various materials, and the methods of preventing them, or of making repairs and preventing further slips after they have once occurred. The subject is treated comprehensively, and contains many notes of practical value, the result of twenty-five years' experience."

THE BUILDER.

"We gladly welcome Mr. Newman's book on slips in earthworks as an important contribution to a right comprehension of such matters.

"There is much in this book that will certainly guard designers of engineering works against probable, if not against possible, slips in earthworks.

"The capital cost of a work and the cost of its maintenance may both be very sensibly reduced by attention to all the points alluded to by the author.

"We are glad to see that the author enters at some length into the subject of the due provision of drainage at the backs of retaining walls, a matter so often neglected or overlooked, and carries this subject to a far larger one, the causes which tend to disturb the repose of dock walls. His remarks on these matters are well worthy of consideration, and are thoroughly practical, and the items which have to be taken into account in the necessary statical calculations very well introduced.

"In conclusion, we may say that there is plenty of good useful information to be obtained from this work, which touches a subject possessing an exceedingly scanty vocabulary.

"It contains an immense deal of matter which must be swallowed sooner or later by every one who desires to be a good engineer."

BUILDING NEWS.

"Mr. John Newman, Assoc.M.Inst.C.E., has written a volume on a subject that has hitherto only been treated of cursorily.

"Useful advice is given, which the railway engineer and earthwork contractor may profit by.

"The book contains a fund of useful information."

BUILDERS' REPORTER AND ENGINEERING TIMES.

"The book which Mr. John Newman has written imparts a new interest to earthworks. It is, in fact, a sort of pathological treatise, and as such may be said to be unique among books on construction, for in them failures are rarely recognised. Now in Mr. Newman's volume the majority of the pages relate to failures, and from them the reader infers how they are to be avoided, and thus to form earthworks that will endure longer than those which are executed without much regard to risks.

"The manner of dealing with the subsidences when they occur, as well as providing against them, will be found described in the book.

"It can be said that the subject is thoroughly investigated, and contractors as well as engineers can learn much from Mr. Newman's book."

E. & F. N. SPON, 125, STRAND, LONDON.

NOTES ON CONCRETE AND WORKS IN CONCRETE.

By JOHN NEWMAN,
Assoc.M.Inst.C.E.

REVIEWS OF THE PRESS.

FIRST EDITION.

Engineering.

"An epitome of the best practice, which may be relied upon not to mislead.

"The successful construction of works in concrete is a difficult matter to explain in books.

"All the points which open the way to bad work are carefully pointed out."

Iron.

"As numerous examples are cited of the use of concrete in public works, and details supplied, *the book will greatly assist engineers engaged upon such works.*"

The Builder.

"A very practical little book, carefully compiled, and *one which all writers of specifications for concrete work would do well to peruse.*

"*The book contains reliable information for all engaged upon public works.*

"A perusal of Mr. Newman's valuable little handbook will point out the importance of a more careful investigation of the subject than is usually supposed to be necessary."

AMERICAN PRESS.

Building.

"To accomplish so much in so limited a space, the subject-matter has been confined to chapters.

"*We take pleasure in saying that this is the most admirable and complete handbook on concretes for engineers of which we have knowledge.*"

E. & F. N. SPON, 125, STRAND, LONDON.

1893.

BOOKS RELATING
TO
APPLIED SCIENCE

PUBLISHED BY

E. & F. N. SPON.

LONDON: 125 STRAND.

NEW YORK: 12 CORTLANDT STREET.

───◆───

The Engineers' Sketch-Book of Mechanical Movements, Devices, Appliances, Contrivances, Details employed in the Design and Construction of Machinery for every purpose. Collected from numerous Sources and from Actual Work. Classified and Arranged for Reference. *Nearly* 2000 *Illustrations*. By T. B. BARBER, Engineer. Second Edition, 8vo, cloth, 7s. 6d.

A Pocket-Book for Chemists, Chemical Manufacturers, Metallurgists, Dyers, Distillers, Brewers, Sugar Refiners, Photographers, Students, etc., etc. By THOMAS BAYLEY, Assoc. R.C. Sc. Ireland, Analytical and Consulting Chemist and Assayer. Fifth edition, 481 pp., royal 32mo, roan, gilt edges, 5s.

SYNOPSIS OF CONTENTS :

Atomic Weights and Factors—Useful Data—Chemical Calculations—Rules for Indirect Analysis—Weights and Measures—Thermometers and Barometers—Chemical Physics—Boiling Points, etc.—Solubility of Substances—Methods of Obtaining Specific Gravity—Conversion of Hydrometers—Strength of Solutions by Specific Gravity—Analysis—Gas Analysis—Water Analysis—Qualitative Analysis and Reactions—Volumetric Analysis—Manipulation—Mineralogy—Assaying—Alcohol—Beer—Sugar—Miscellaneous Technological matter relating to Potash, Soda, Sulphuric Acid, Chlorine, Tar Products, Petroleum, Milk, Tallow, Photography, Prices, Wages, Appendix, etc., etc.

Electricity, its Theory, Sources, and Applications.
By JOHN T. SPRAGUE, M. Inst. E.E. Third edition, thoroughly revised and extended, *with numerous illustrations and tables*, crown 8vo, cloth, 15s.

Electric Toys. Electric Toy-Making, Dynamo Building and Electric Motor Construction for Amateurs. By T. O'CONOR SLOANE, Ph.D. *With cuts*, square 16mo, cloth, 4s. 6d.

The Phonograph, and How to Construct it. With a Chapter on Sound. By W. GILLETT. *With engravings and full working drawings*, crown 8vo, cloth, 5s.

Just Published, in Demy 8vo, cloth, containing 975 pages and 250 Illustrations, price 7s. 6d.

SPONS' HOUSEHOLD MANUAL:
A Treasury of Domestic Receipts and Guide for Home Management.

PRINCIPAL CONTENTS.

Hints for selecting a good House, pointing out the essential requirements for a good house as to the Site, Soil, Trees, Aspect, Construction, and General Arrangement; with instructions for Reducing Echoes, Waterproofing Damp Walls, Curing Damp Cellars.

Sanitation.—What should constitute a good Sanitary Arrangement; Examples (with Illustrations) of Well- and Ill-drained Houses; How to Test Drains; Ventilating Pipes, etc.

Water Supply.—Care of Cisterns; Sources of Supply; Pipes; Pumps; Purification and Filtration of Water.

Ventilation and Warming.—Methods of Ventilating without causing cold draughts, by various means; Principles of Warming; Health Questions; Combustion; Open Grates; Open Stoves; Fuel Economisers; Varieties of Grates; Close-Fire Stoves; Hot-air Furnaces; Gas Heating; Oil Stoves; Steam Heating; Chemical Heaters; Management of Flues; and Cure of Smoky Chimneys.

Lighting.—The best methods of Lighting; Candles, Oil Lamps, Gas, Incandescent Gas, Electric Light; How to test Gas Pipes; Management of Gas.

Furniture and Decoration.—Hints on the Selection of Furniture; on the most approved methods of Modern Decoration; on the best methods of arranging Bells and Calls; How to Construct an Electric Bell.

Thieves and Fire.—Precautions against Thieves and Fire; Methods of Detection; Domestic Fire Escapes; Fireproofing Clothes, etc.

The Larder.—Keeping Food fresh for a limited time; Storing Food without change, such as Fruits, Vegetables, Eggs, Honey, etc.

Curing Foods for lengthened Preservation, as Smoking, Salting, Canning, Potting, Pickling, Bottling Fruits, etc.; Jams, Jellies, Marmalade, etc.

The Dairy.—The Building and Fitting of Dairies in the most approved modern style; Butter-making; Cheesemaking and Curing.

The Cellar.—Building and Fitting; Cleaning Casks and Bottles; Corks and Corking; Aërated Drinks; Syrups for Drinks; Beers; Bitters; Cordials and Liqueurs; Wines; Miscellaneous Drinks.

The Pantry.—Bread-making; Ovens and Pyrometers; Yeast; German Yeast; Biscuits; Cakes; Fancy Breads; Buns.

The Kitchen.—On Fitting Kitchens; a description of the best Cooking Ranges, close and open; the Management and Care of Hot Plates, Baking Ovens, Dampers, Flues, and Chimneys; Cooking by Gas; Cooking by Oil; the Arts of Roasting, Grilling, Boiling, Stewing, Braising, Frying.

Receipts for Dishes—Soups, Fish, Meat, Game, Poultry, Vegetables, Salads, Puddings, Pastry, Confectionery, Ices, etc., etc.; Foreign Dishes.

The Housewife's Room.—Testing Air, Water, and Foods; Cleaning and Renovating; Destroying Vermin.

Housekeeping, Marketing.

The Dining-Room.—Dietetics; Laying and Waiting at Table: Carving; Dinners, Breakfasts, Luncheons, Teas, Suppers, etc.

The Drawing-Room.—Etiquette; Dancing; Amateur Theatricals; Tricks and Illusions; Games (indoor).

The Bedroom and Dressing-Room; Sleep; the Toilet; Dress; Buying Clothes; Outfits; Fancy Dress.

The Nursery.—The Room; Clothing; Washing; Exercise; Sleep; Feeding; Teething; Illness; Home Training.

The Sick-Room.—The Room; the Nurse; the Bed; Sick Room Accessories; Feeding Patients; Invalid Dishes and Drinks; Administering Physic; Domestic Remedies; Accidents and Emergencies; Bandaging; Burns; Carrying Injured Persons; Wounds; Drowning; Fits; Frost-bites; Poisons and Antidotes; Sunstroke; Common Complaints; Disinfection, etc.

The Bath-Room.—Bathing in General; Management of Hot-Water System.
The Laundry.—Small Domestic Washing Machines, and methods of getting up linen.
Fitting up and Working a Steam Laundry.
The School-Room.—The Room and its Fittings; Teaching, etc.
The Playground.—Air and Exercise; Training; Outdoor Games and Sports.
The Workroom.—Darning, Patching, and Mending Garments.
The Library.—Care of Books.
The Garden.—Calendar of Operations for Lawn, Flower Garden, and Kitchen Garden.
The Farmyard.—Management of the Horse, Cow, Pig, Poultry, Bees, etc., etc.
Small Motors.—A description of the various small Engines useful for domestic purposes, from 1 man to 1 horse power, worked by various methods, such as Electric Engines, Gas Engines, Petroleum Engines, Steam Engines, Condensing Engines, Water Power, Wind Power, and the various methods of working and managing them.
Household Law.—The Law relating to Landlords and Tenants, Lodgers, Servants, Parochial Authorities, Juries, Insurance, Nuisance, etc.

On Designing Belt Gearing. By E. J. COWLING WELCH, Mem. Inst. Mech. Engineers, Author of 'Designing Valve Gearing.' Fcap. 8vo, sewed, 6*d.*

A Handbook of Formulæ, Tables, and Memoranda, for Architectural Surveyors and others engaged in Building. By J. T. HURST, C.E. Fourteenth edition, royal 32mo, roan, 5*s.*

"It is no disparagement to the many excellent publications we refer to, to say that in our opinion this little pocket-book of Hurst's is the very best of them all, without any exception. It would be useless to attempt a recapitulation of the contents, for it appears to contain almost *everything* that anyone connected with building could require, and, best of all, made up in a compact form for carrying in the pocket, measuring only 5 in. by 3 in., and about ¼ in. thick, in a limp cover. We congratulate the author on the success of his laborious and practically compiled little book, which has received unqualified and deserved praise from every professional person to whom we have shown it."—*The Dublin Builder.*

Tabulated Weights of Angle, Tee, Bulb, Round, Square, and Flat Iron and Steel, and other information for the use of Naval Architects and Shipbuilders. By C. H. JORDAN, M.I.N.A. Fourth edition, 32mo, cloth, 2*s.* 6*d.*

A Complete Set of Contract Documents for a Country Lodge, comprising Drawings, Specifications, Dimensions (for quantities), Abstracts, Bill of Quantities, Form of Tender and Contract, with Notes by J. LEANING, printed in facsimile of the original documents, on single sheets fcap., in paper case, 10*s.*

A Practical Treatise on Heat, as applied to the Useful Arts; for the Use of Engineers, Architects, &c. By THOMAS BOX. With 14 *plates.* Sixth edition, crown 8vo, cloth, 12*s.* 6*d.*

A Descriptive Treatise on Mathematical Drawing Instruments: their construction, uses, qualities, selection, preservation, and suggestions for improvements, with hints upon Drawing and Colouring. By W. F. STANLEY, M.R.I. Sixth edition, *with numerous illustrations,* crown 8vo, cloth, 5*s.*

B 2

Quantity Surveying. By J. LEANING. With 42 illustrations. Second edition, revised, crown 8vo, cloth, 9s.

CONTENTS:

A complete Explanation of the London Practice.
General Instructions.
Order of Taking Off.
Modes of Measurement of the various Trades.
Use and Waste.
Ventilation and Warming.
Credits, with various Examples of Treatment.
Abbreviations.
Squaring the Dimensions.
Abstracting, with Examples in illustration of each Trade.
Billing.
Examples of Preambles to each Trade.
Form for a Bill of Quantities.
Do. Bill of Credits.
Do. Bill for Alternative Estimate.
Restorations and Repairs, and Form of Bill.
Variations before Acceptance of Tender.
Errors in a Builder's Estimate.

Schedule of Prices.
Form of Schedule of Prices.
Analysis of Schedule of Prices.
Adjustment of Accounts.
Form of a Bill of Variations.
Remarks on Specifications.
Prices and Valuation of Work, with Examples and Remarks upon each Trade.
The Law as it affects Quantity Surveyors, with Law Reports.
Taking Off after the Old Method.
Northern Practice.
The General Statement of the Methods recommended by the Manchester Society of Architects for taking Quantities.
Examples of Collections.
Examples of "Taking Off" in each Trade.
Remarks on the Past and Present Methods of Estimating.

Spons' Architects' and Builders' Price Book, with useful *Memoranda.* Edited by W. YOUNG, Architect. Crown 8vo, cloth, red edges, 3s. 6d. *Published annually.* Nineteenth edition. *Now ready.*

Long-Span Railway Bridges, comprising Investigations of the Comparative Theoretical and Practical Advantages of the various adopted or proposed Type Systems of Construction, with numerous Formulæ and Tables giving the weight of Iron or Steel required in Bridges from 300 feet to the limiting Spans; to which are added similar Investigations and Tables relating to Short-span Railway Bridges. Second and revised edition. By B. BAKER, Assoc. Inst. C.E. *Plates,* crown 8vo, cloth, 5s.

Elementary Theory and Calculation of Iron Bridges and *Roofs.* By AUGUST RITTER, Ph.D., Professor at the Polytechnic School at Aix-la-Chapelle. Translated from the third German edition, by H. R. SANKEY, Capt. R.E. With 500 *illustrations,* 8vo, cloth, 15s.

The Elementary Principles of Carpentry. By THOMAS TREDGOLD. Revised from the original edition, and partly re-written, by JOHN THOMAS HURST. Contained in 517 pages of letterpress, and *illustrated with* 48 *plates and* 150 *wood engravings.* Sixth edition, reprinted from the third, crown 8vo, cloth, 12s. 6d.

Section I. On the Equality and Distribution of Forces—Section II. Resistance of Timber—Section III. Construction of Floors—Section IV. Construction of Roofs—Section V. Construction of Domes and Cupolas—Section VI. Construction of Partitions—Section VII. Scaffolds, Staging, and Gantries—Section VIII. Construction of Centres for Bridges—Section IX. Coffer-dams, Shoring, and Strutting—Section X. Wooden Bridges and Viaducts—Section XI. Joints, Straps, and other Fastenings—Section XII. Timber.

The Builder's Clerk: a Guide to the Management of a Builder's Business. By THOMAS BALES. Fcap. 8vo, cloth, 1s. 6d.

Practical Gold-Mining: a Comprehensive Treatise on the Origin and Occurrence of Gold-bearing Gravels, Rocks and Ores, and the methods by which the Gold is extracted. By C. G. WARNFORD LOCK, co-Author of 'Gold: its Occurrence and Extraction.' *With* 8 *plates and* 275 *engravings in the text*, royal 8vo, cloth, 2*l*. 2*s*.

Hot Water Supply: A Practical Treatise upon the Fitting of Circulating Apparatus in connection with Kitchen Range and other Boilers, to supply Hot Water for Domestic and General Purposes. With a Chapter upon Estimating. *Fully illustrated*, crown 8vo, cloth, 3*s*.

Hot Water Apparatus: An Elementary Guide for the Fitting and Fixing of Boilers and Apparatus for the Circulation of Hot Water for Heating and for Domestic Supply, and containing a Chapter upon Boilers and Fittings for Steam Cooking. 32 *illustrations*, fcap. 8vo, cloth, 1*s*. 6*d*.

The Use and Misuse, and the Proper and Improper Fixing of a Cooking Range. *Illustrated*, fcap. 8vo, sewed, 6*d*.

Iron Roofs: Examples of Design, Description. Illustrated with 64 *Working Drawings of Executed Roofs.* By ARTHUR T. WALMISLEY, Assoc. Mem. Inst. C.E. Second edition, revised, imp. 4to, half-morocco, 3*l*. 3*s*.

A History of Electric Telegraphy, to the Year 1837. Chiefly compiled from Original Sources, and hitherto Unpublished Documents, by J. J. FAHIE, Mem. Soc. of Tel. Engineers, and of the International Society of Electricians, Paris. Crown 8vo, cloth, 9*s*.

Spons' Information for Colonial Engineers. Edited by J. T. HURST. Demy 8vo, sewed.

No. 1, Ceylon. By ABRAHAM DEANE, C.E. 2*s*. 6*d*.

CONTENTS:
Introductory Remarks—Natural Productions—Architecture and Engineering—Topography, Trade, and Natural History—Principal Stations—Weights and Measures, etc., etc.

No. 2. Southern Africa, including the Cape Colony, Natal, and the Dutch Republics. By HENRY HALL, F.R.G.S., F.R.C.I. With Map. 3*s*. 6*d*.

CONTENTS:
General Description of South Africa—Physical Geography with reference to Engineering Operations—Notes on Labour and Material in Cape Colony—Geological Notes on Rock Formation in South Africa—Engineering Instruments for Use in South Africa—Principal Public Works in Cape Colony: Railways, Mountain Roads and Passes, Harbour Works, Bridges, Gas Works, Irrigation and Water Supply, Lighthouses, Drainage and Sanitary Engineering, Public Buildings, Mines—Table of Woods in South Africa—Animals used for Draught Purposes—Statistical Notes—Table of Distances—Rates of Carriage, etc.

No. 3. India. By F. C. DANVERS, Assoc. Inst. C.E. With Map. 4*s*. 6*d*.

CONTENTS:
Physical Geography of India—Building Materials—Roads—Railways—Bridges—Irrigation—River Works—Harbours—Lighthouse Buildings—Native Labour—The Principal Trees of India—Money—Weights and Measures—Glossary of Indian Terms, etc.

Our Factories, Workshops, and Warehouses: their Sanitary and Fire-Resisting Arrangements. By B. H. THWAITE, Assoc. Mem. Inst. C.E. With 183 *wood engravings*, crown 8vo, cloth, 9s.

A Practical Treatise on Coal Mining. By GEORGE G. ANDRÉ, F.G.S., Assoc. Inst. C.E., Member of the Society of Engineers. With 82 *lithographic plates*. 2 vols., royal 4to, cloth, 3l. 12s.

A Practical Treatise on Casting and Founding, including descriptions of the modern machinery employed in the art. By N. E. SPRETSON, Engineer. Fifth edition, with 82 *plates* drawn to scale, 412 pp., demy 8vo, cloth, 18s.

A Handbook of Electrical Testing. By H. R. KEMPE, M.S.T.E. Fourth edition, revised and enlarged, crown 8vo, cloth, 16s.

The Clerk of Works: a Vade-Mecum for all engaged in the Superintendence of Building Operations. By G. G. HOSKINS, F.R.I.B.A. Third edition, fcap. 8vo, cloth, 1s. 6d.

American Foundry Practice: Treating of Loam, Dry Sand, and Green Sand Moulding, and containing a Practical Treatise upon the Management of Cupolas, and the Melting of Iron. By T. D. WEST, Practical Iron Moulder and Foundry Foreman. Second edition, *with numerous illustrations,* crown 8vo, cloth, 10s. 6d.

The Maintenance of Macadamised Roads. By T. CODRINGTON, M.I.C.E, F.G.S., General Superintendent of County Roads for South Wales. Second edition, 8vo, cloth, 7s. 6d.

Hydraulic Steam and Hand Power Lifting and Pressing Machinery. By FREDERICK COLYER, M. Inst. C.E., M. Inst. M.E. With 73 *plates*, 8vo, cloth, 18s.

Pumps and Pumping Machinery. By F. COLYER, M.I.C.E., M.I.M.E. With 23 *folding plates*, 8vo, cloth, 12s. 6d.

Pumps and Pumping Machinery. By F. COLYER. Second Part. With 11 *large plates*, 8vo, cloth, 12s. 6d.

A Treatise on the Origin, Progress, Prevention, and Cure of Dry Rot in Timber; with Remarks on the Means of Preserving Wood from Destruction by Sea-Worms, Beetles, Ants, etc. By THOMAS ALLEN BRITTON, late Surveyor to the Metropolitan Board of Works, etc., etc. With 10 *plates*, crown 8vo, cloth, 7s. 6d.

The Artillery of the Future and the New Powders. By J. A. LONGRIDGE, Mem. Inst. C.E. 8vo, cloth, 5s.

Gas Works: their Arrangement, Construction, Plant,
and Machinery. By F. COLYER, M. Inst. C.E. *With 31 folding plates,*
8vo, cloth, 12s. 6d.

The Municipal and Sanitary Engineer's Handbook.
By H. PERCY BOULNOIS, Mem. Inst. C.E., Borough Engineer, Portsmouth. *With numerous illustrations.* Second edition, demy 8vo, cloth, 15s.

CONTENTS:

The Appointment and Duties of the Town Surveyor—Traffic—Macadamised Roadways—Steam Rolling—Road Metal and Breaking—Pitched Pavements—Asphalte—Wood Pavements—Footpaths—Kerbs and Gutters—Street Naming and Numbering—Street Lighting—Sewerage—Ventilation of Sewers—Disposal of Sewage—House Drainage—Disinfection—Gas and Water Companies, etc., Breaking up Streets—Improvement of Private Streets—Borrowing Powers—Artizans' and Labourers' Dwellings—Public Conveniences—Scavenging, including Street Cleansing—Watering and the Removing of Snow—Planting Street Trees—Deposit of Plans—Dangerous Buildings—Hoardings—Obstructions—Improving Street Lines—Cellar Openings—Public Pleasure Grounds—Cemeteries—Mortuaries—Cattle and Ordinary Markets —Public Slaughter-houses, etc.—Giving numerous Forms of Notices, Specifications, and General Information upon these and other subjects of great importance to Municipal Engineers and others engaged in Sanitary Work.

Metrical Tables. By Sir G. L. MOLESWORTH,
M.I.C.E. 32mo, cloth, 1s. 6d.

CONTENTS.

General—Linear Measures—Square Measures—Cubic Measures—Measures of Capacity—Weights—Combinations—Thermometers.

Elements of Construction for Electro-Magnets. By
Count TH. DU MONCEL, Mem. de l'Institut de France. Translated from the French by C. J. WHARTON. Crown 8vo, cloth, 4s. 6d.

A Treatise on the Use of Belting for the Transmission of Power. By J. H. COOPER. Second edition, *illustrated,* 8vo, cloth, 15s.

A Pocket-Book of Useful Formulæ and Memoranda
for Civil and Mechanical Engineers. By Sir GUILFORD L. MOLESWORTH, Mem. Inst. C.E. *With numerous illustrations,* 744 pp. Twenty-second edition, 32mo, roan, 6s.

SYNOPSIS OF CONTENTS:

Surveying, Levelling, etc.—Strength and Weight of Materials—Earthwork, Brickwork, Masonry, Arches, etc.—Struts, Columns, Beams, and Trusses—Flooring, Roofing, and Roof Trusses—Girders, Bridges, etc.—Railways and Roads—Hydraulic Formulæ—Canals, Sewers, Waterworks, Docks—Irrigation and Breakwaters—Gas, Ventilation, and Warming—Heat, Light, Colour, and Sound—Gravity: Centres, Forces, and Powers—Millwork, Teeth of Wheels, Shafting, etc.—Workshop Recipes—Sundry Machinery—Animal Power—Steam and the Steam Engine—Water-power, Water-wheels, Turbines, etc.—Wind and Windmills—Steam Navigation, Ship Building, Tonnage, etc.—Gunnery, Projectiles, etc.—Weights, Measures, and Money—Trigonometry, Conic Sections, and Curves—Telegraphy—Mensuration—Tables of Areas and Circumference, and Arcs of Circles—Logarithms, Square and Cube Roots, Powers—Reciprocals, etc.—Useful Numbers—Differential and Integral Calculus—Algebraic Signs—Telegraphic Construction and Formulæ.

Hints on Architectural Draughtsmanship. By G. W. TUXFORD HALLATT. Fcap. 8vo, cloth, 1s. 6d.

Spons' Tables and Memoranda for Engineers; selected and arranged by J. T. HURST, C.E., Author of 'Architectural Surveyors' Handbook,' 'Hurst's Tredgold's Carpentry,' etc. Eleventh edition, 64mo, roan, gilt edges, 1s.; or in cloth case, 1s. 6d.

This work is printed in a pearl type, and is so small, measuring only 2½ in. by 1¾ in. by ¼ in. thick, that it may be easily carried in the waistcoat pocket.

"It is certainly an extremely rare thing for a reviewer to be called upon to notice a volume measuring but 2½ in. by 1¾ in., yet these dimensions faithfully represent the size of the handy little book before us. The volume—which contains 118 printed pages, besides a few blank pages for memoranda—is, in fact, a true pocket-book, adapted for being carried in the waistcoat pocket, and containing a far greater amount and variety of information than most people would imagine could be compressed into so small a space. . . . . The little volume has been compiled with considerable care and judgment, and we can cordially recommend it to our readers as a useful little pocket companion."—*Engineering.*

A Practical Treatise on Natural and Artificial Concrete, its Varieties and Constructive Adaptations. By HENRY REID, Author of the 'Science and Art of the Manufacture of Portland Cement.' New Edition, *with* 59 *woodcuts and* 5 *plates,* 8vo, cloth, 15s.

Notes on Concrete and Works in Concrete; especially written to assist those engaged upon Public Works. By JOHN NEWMAN, Assoc. Mem. Inst. C.E., crown 8vo, cloth, 4s. 6d.

Electricity as a Motive Power. By Count TH. DU MONCEL, Membre de l'Institut de France, and FRANK GERALDY, Ingénieur des Ponts et Chaussées. Translated and Edited, with Additions, by C. J. WHARTON, Assoc. Soc. Tel. Eng. and Elec. With 113 *engravings and diagrams,* crown 8vo, cloth, 7s. 6d.

Treatise on Valve-Gears, with special consideration of the Link-Motions of Locomotive Engines. By Dr. GUSTAV ZEUNER, Professor of Applied Mechanics at the Confederated Polytechnikum of Zurich. Translated from the Fourth German Edition, by Professor J. F. KLEIN, Lehigh University, Bethlehem, Pa. *Illustrated,* 8vo, cloth, 12s. 6d.

The French-Polisher's Manual. By a French-Polisher; containing Timber Staining, Washing, Matching, Improving, Painting, Imitations, Directions for Staining, Sizing, Embodying, Smoothing, Spirit Varnishing, French-Polishing, Directions for Repolishing. Third edition, royal 32mo, sewed, 6d.

Hops, their Cultivation, Commerce, and Uses in various Countries. By P. L. SIMMONDS. Crown 8vo, cloth, 4s. 6d.

The Principles of Graphic Statics. By GEORGE SYDENHAM CLARKE, Major Royal Engineers. *With* 112 *illustrations.* Second edition, 4to, cloth, 12s. 6d.

PUBLISHED BY E. & F. N. SPON. 9

Dynamo Tenders' Hand-Book. By F. B. BADT, late
1st Lieut. Royal Prussian Artillery. With 70 *illustrations.* Third edition,
18mo, cloth, 4s. 6d.

*Practical Geometry, Perspective, and Engineering
Drawing;* a Course of Descriptive Geometry adapted to the Requirements of the Engineering Draughtsman, including the determination of cast shadows and Isometric Projection, each chapter being followed by numerous examples ; to which are added rules for Shading, Shade-lining, etc., together with practical instructions as to the Lining, Colouring, Printing, and general treatment of Engineering Drawings, with a chapter on drawing Instruments. By GEORGE S. CLARKE, Capt. R.E. Second edition, *with* 21 *plates.* 2 vols., cloth, 10s. 6d.

The Elements of Graphic Statics. By Professor
KARL VON OTT, translated from the German by G. S. CLARKE, Capt. R.E., Instructor in Mechanical Drawing, Royal Indian Engineering College. *With* 93 *illustrations,* crown 8vo, cloth, 5s.

A Practical Treatise on the Manufacture and Distribution of Coal Gas. By WILLIAM RICHARDS. Demy 4to, with *numerous wood engravings and* 29 *plates,* cloth, 28s.

SYNOPSIS OF CONTENTS :

Introduction — History of Gas Lighting — Chemistry of Gas Manufacture, by Lewis Thompson, Esq., M.R.C.S.—Coal, with Analyses, by J. Paterson, Lewis Thompson, and G. R. Hislop, Esqrs.—Retorts, Iron and Clay—Retort Setting—Hydraulic Main—Condensers — Exhausters — Washers and Scrubbers — Purifiers — Purification — History of Gas Holder — Tanks, Brick and Stone, Composite, Concrete, Cast-iron, Compound Annular Wrought-iron — Specifications — Gas Holders — Station Meter — Governor — Distribution— Mains—Gas Mathematics, or Formulæ for the Distribution of Gas, by Lewis Thompson, Esq.— Services—Consumers' Meters—Regulators—Burners—Fittings—Photometer—Carburization of Gas—Air Gas and Water Gas—Composition of Coal Gas, by Lewis Thompson, Esq.— Analyses of Gas—Influence of Atmospheric Pressure and Temperature on Gas—Residual Products—Appendix—Description of Retort Settings, Buildings, etc., etc.

*The New Formula for Mean Velocity of Discharge
of Rivers and Canals.* By W. R. KUTTER. Translated from articles in the 'Cultur-Ingénieur,' by LOWIS D'A. JACKSON, Assoc. Inst. C.E. 8vo, cloth, 12s. 6d.

*The Practical Millwright and Engineer's Ready
Reckoner;* or Tables for finding the diameter and power of cog-wheels, diameter, weight, and power of shafts, diameter and strength of bolts, etc. By THOMAS DIXON. Fourth edition, 12mo, cloth, 3s.

Tin: Describing the Chief Methods of Mining,
Dressing and Smelting it abroad ; with Notes upon Arsenic, Bismuth and Wolfram. By ARTHUR G. CHARLETON, Mem. American Inst. of Mining Engineers. *With plates,* 8vo, cloth, 12s. 6d.

B 3

Perspective, Explained and Illustrated. By G. S. CLARKE, Capt. R.E. *With illustrations,* 8vo, cloth, 3s. 6d.

Practical Hydraulics; a Series of Rules and Tables for the use of Engineers, etc., etc. By THOMAS BOX. Ninth edition, *numerous plates,* post 8vo, cloth, 5s.

The Essential Elements of Practical Mechanics; based on the Principle of Work, designed for Engineering Students. By OLIVER BYRNE, formerly Professor of Mathematics, College for Civil Engineers. Third edition, *with* 148 *wood engravings,* post 8vo, cloth, 7s. 6d.

CONTENTS:

Chap. 1. How Work is Measured by a Unit, both with and without reference to a Unit of Time—Chap. 2. The Work of Living Agents, the Influence of Friction, and introduces one of the most beautiful Laws of Motion—Chap. 3. The principles expounded in the first and second chapters are applied to the Motion of Bodies—Chap. 4. The Transmission of Work by simple Machines—Chap. 5. Useful Propositions and Rules.

Breweries and Maltings: their Arrangement, Construction, Machinery, and Plant. By G. SCAMELL, F.R.I.B.A. Second edition, revised, enlarged, and partly rewritten. By F. COLYER, M.I.C.E., M.I.M.E. *With* 20 *plates,* 8vo, cloth, 12s. 6d.

A Practical Treatise on the Construction of Horizontal and Vertical Waterwheels, specially designed for the use of operative mechanics. By WILLIAM CULLEN, Millwright and Engineer. *With* 11 *plates.* Second edition, revised and enlarged, small 4to, cloth, 12s. 6d.

A Practical Treatise on Mill-gearing, Wheels, Shafts, Riggers, etc.; for the use of Engineers. By THOMAS BOX. Third edition, *with* 11 *plates.* Crown 8vo, cloth, 7s. 6d.

Mining Machinery: a Descriptive Treatise on the Machinery, Tools, and other Appliances used in Mining. By G. G. ANDRÉ, F.G.S., Assoc. Inst. C.E., Mem. of the Society of Engineers. Royal 4to, uniform with the Author's Treatise on Coal Mining, containing 182 *plates,* accurately drawn to scale, with descriptive text, in 2 vols., cloth, 3l. 12s.

CONTENTS:

Machinery for Prospecting, Excavating, Hauling, and Hoisting—Ventilation—Pumping—Treatment of Mineral Products, including Gold and Silver, Copper, Tin, and Lead, Iron, Coal, Sulphur, China Clay, Brick Earth, etc.

Tables for Setting out Curves for Railways, Canals, Roads, etc., varying from a radius of five chains to three miles. By A. KENNEDY and R. W. HACKWOOD. *Illustrated* 32mo, cloth, 2s. 6d.

Practical Electrical Notes and Definitions for the
use of Engineering Students and Practical Men. By W. PERREN
MAYCOCK, Assoc. M. Inst. E.E., Instructor in Electrical Engineering at
the Pitlake Institute, Croydon, together with the Rules and Regulations
to be observed in Electrical Installation Work. Second edition. Royal
32mo, roan, gilt edges, 4s. 6d., or cloth, red edges, 3s.

The Draughtsman's Handbook of Plan and Map
Drawing; including instructions for the preparation of Engineering,
Architectural, and Mechanical Drawings. *With numerous illustrations
in the text, and* 33 *plates* (15 *printed in colours*). By G. G. ANDRÉ,
F.G.S., Assoc. Inst. C.E. 4to, cloth, 9s.

CONTENTS:

The Drawing Office and its Furnishings—Geometrical Problems—Lines, Dots, and their Combinations—Colours, Shading, Lettering, Bordering, and North Points—Scales—Plotting—Civil Engineers' and Surveyors' Plans—Map Drawing—Mechanical and Architectural Drawing—Copying and Reducing Trigonometrical Formulæ, etc., etc.

The Boiler-maker's and Iron Ship-builder's Companion,
comprising a series of original and carefully calculated tables, of the
utmost utility to persons interested in the iron trades. By JAMES FODEN,
author of 'Mechanical Tables,' etc. Second edition revised, *with illustrations*, crown 8vo, cloth, 5s.

Rock Blasting: a Practical Treatise on the means
employed in Blasting Rocks for Industrial Purposes. By G. G. ANDRÉ,
F.G.S., Assoc. Inst. C.E. With 56 *illustrations and* 12 *plates*, 8vo, cloth,
10s. 6d.

Experimental Science: Elementary, Practical, and
Experimental Physics. By GEO. M. HOPKINS. *Illustrated by* 672
engravings. In one large vol., 8vo, cloth, 15s.

A Treatise on Ropemaking as practised in public and
private Rope-yards, with a Description of the Manufacture, Rules, Tables
of Weights, etc., adapted to the Trade, Shipping, Mining, Railways,
Builders, etc. By R. CHAPMAN, formerly foreman to Messrs. Huddart
and Co., Limehouse, and late Master Ropemaker to H.M. Dockyard,
Deptford. Second edition, 12mo, cloth, 3s.

Laxton's Builders' and Contractors' Tables; for the
use of Engineers, Architects, Surveyors, Builders, Land Agents, and
others. Bricklayer, containing 22 tables, with nearly 30,000 calculations.
4to, cloth, 5s.

Laxton's Builders' and Contractors' Tables. Excavator, Earth, Land, Water, and Gas, containing 53 tables, with nearly
24,000 calculations. 4to, cloth, 5s.

Egyptian Irrigation. By W. WILLCOCKS, M.I.C.E., Indian Public Works Department, Inspector of Irrigation, Egypt. With Introduction by Lieut.-Col. J. C. ROSS, R.E., Inspector-General of Irrigation. *With numerous lithographs and wood engravings*, royal 8vo, cloth, 1*l*. 16*s*.

Screw Cutting Tables for Engineers and Machinists, giving the values of the different trains of Wheels required to produce Screws of any pitch, calculated by Lord Lindsay, M.P., F.R.S., F.R.A.S., etc. Cloth, oblong, 2*s*.

Screw Cutting Tables, for the use of Mechanical Engineers, showing the proper arrangement of Wheels for cutting the Threads of Screws of any required pitch, with a Table for making the Universal Gas-pipe Threads and Taps. By W. A. MARTIN, Engineer. Second edition, oblong, cloth, 1*s*., or sewed, 6*d*.

A Treatise on a Practical Method of Designing Slide-Valve Gears by Simple Geometrical Construction, based upon the principles enunciated in Euclid's Elements, and comprising the various forms of Plain Slide-Valve and Expansion Gearing; together with Stephenson's, Gooch's, and Allan's Link-Motions, as applied either to reversing or to variable expansion combinations. By EDWARD J. COWLING WELCH, Memb. Inst. Mechanical Engineers. Crown 8vo, cloth, 6*s*.

Cleaning and Scouring: a Manual for Dyers, Laundresses, and for Domestic Use. By S. CHRISTOPHER. 18mo, sewed, 6*d*.

A Glossary of Terms used in Coal Mining. By WILLIAM STUKELEY GRESLEY, Assoc. Mem. Inst. C.E., F.G.S., Member of the North of England Institute of Mining Engineers. *Illustrated with numerous woodcuts and diagrams*, crown 8vo, cloth, 5*s*.

A Pocket-Book for Boiler Makers and Steam Users, comprising a variety of useful information for Employer and Workman, Government Inspectors, Board of Trade Surveyors, Engineers in charge of Works and Slips, Foremen of Manufactories, and the general Steam-using Public. By MAURICE JOHN SEXTON. Second edition, royal 32mo, roan, gilt edges, 5*s*.

Electrolysis: a Practical Treatise on Nickeling, Coppering, Gilding, Silvering, the Refining of Metals, and the treatment of Ores by means of Electricity. By HIPPOLYTE FONTAINE, translated from the French by J. A. BERLY, C.E., Assoc. S.T.E. *With engravings*. 8vo, cloth, 9*s*.

Barlow's Tables of Squares, Cubes, Square Roots,
Cube Roots, Reciprocals of all Integer Numbers up to 10,000. Post 8vo, cloth, 6s.

A Practical Treatise on the Steam Engine, containing Plans and Arrangements of Details for Fixed Steam Engines, with Essays on the Principles involved in Design and Construction. By ARTHUR RIGG, Engineer, Member of the Society of Engineers and of the Royal Institution of Great Britain. Demy 4to, *copiously illustrated with woodcuts and 96 plates,* in one Volume, half-bound morocco, 2l. 2s.; or cheaper edition, cloth, 25s.

This work is not, in any sense, an elementary treatise, or history of the steam engine, but is intended to describe examples of Fixed Steam Engines without entering into the wide domain of locomotive or marine practice. To this end illustrations will be given of the most recent arrangements of Horizontal, Vertical, Beam, Pumping, Winding, Portable, Semi-portable, Corliss, Allen, Compound, and other similar Engines, by the most eminent Firms in Great Britain and America. The laws relating to the action and precautions to be observed in the construction of the various details, such as Cylinders, Pistons, Piston-rods, Connecting-rods, Cross-heads, Motion-blocks, Eccentrics, Simple, Expansion, Balanced, and Equilibrium Slide-valves, and Valve-gearing will be minutely dealt with. In this connection will be found articles upon the Velocity of Reciprocating Parts and the Mode of Applying the Indicator, Heat and Expansion of Steam Governors, and the like. It is the writer's desire to draw illustrations from every possible source, and give only those rules that present practice deems correct.

A Practical Treatise on the Science of Land and Engineering Surveying, Levelling, Estimating Quantities, etc., with a general description of the several Instruments required for Surveying, Levelling, Plotting, etc. By H. S. MERRETT. Fourth edition, revised by G. W. USILL, Assoc. Mem. Inst. C.E. 41 *plates, with illustrations and tables,* royal 8vo, cloth, 12s. 6d.

PRINCIPAL CONTENTS:

Part 1. Introduction and the Principles of Geometry. Part 2. Land Surveying; comprising General Observations—The Chain—Offsets Surveying by the Chain only—Surveying Hilly Ground—To Survey an Estate or Parish by the Chain only—Surveying with the Theodolite—Mining and Town Surveying—Railroad Surveying—Mapping—Division and Laying out of Land—Observations on Enclosures—Plane Trigonometry. Part 3. Levelling—Simple and Compound Levelling—The Level Book—Parliamentary Plan and Section—Levelling with a Theodolite—Gradients—Wooden Curves—To Lay out a Railway Curve—Setting out Widths. Part 4. Calculating Quantities generally for Estimates—Cuttings and Embankments—Tunnels—Brickwork—Ironwork—Timber Measuring. Part 5. Description and Use of Instruments in Surveying and Plotting—The Improved Dumpy Level—Troughton's Level—The Prismatic Compass—Proportional Compass—Box Sextant—Vernier—Pantagraph—Merrett's Improved Quadrant—Improved Computation Scale—The Diagonal Scale—Straight Edge and Sector. Part 6. Logarithms of Numbers—Logarithmic Sines and Co-Sines, Tangents and Co-Tangents—Natural Sines and Co-Sines—Tables for Earthwork, for Setting out Curves, and for various Calculations, etc., etc., etc.

Mechanical Graphics. A Second Course of Mechanical Drawing. With Preface by Prof. PERRY, B.Sc., F.R.S. Arranged for use in Technical and Science and Art Institutes, Schools and Colleges, by GEORGE HALLIDAY, Whitworth Scholar. 8vo, cloth, 6s.

The Assayer's Manual: an Abridged Treatise on the Docimastic Examination of Ores and Furnace and other Artificial Products. By BRUNO KERL. Translated by W. T. BRANNT. *With 65 illustrations*, 8vo, cloth, 12s. 6d.

Dynamo - Electric Machinery: a Text-Book for Students of Electro-Technology. By SILVANUS P. THOMPSON, B.A., D.Sc., M.S.T.E. [*New edition in the press.*

The Practice of Hand Turning in Wood, Ivory, Shell, etc., with Instructions for Turning such Work in Metal as may be required in the Practice of Turning in Wood, Ivory, etc.; also an Appendix on Ornamental Turning. (A book for beginners.) By FRANCIS CAMPIN. Third edition, *with wood engravings*, crown 8vo, cloth, 6s.

CONTENTS:

On Lathes—Turning Tools—Turning Wood—Drilling—Screw Cutting—Miscellaneous Apparatus and Processes—Turning Particular Forms—Staining—Polishing—Spinning Metals —Materials—Ornamental Turning, etc.

Treatise on Watchwork, Past and Present. By the Rev. H. L. NELTHROPP, M.A., F.S.A. *With 32 illustrations*, crown 8vo, cloth, 6s. 6d.

CONTENTS:

Definitions of Words and Terms used in Watchwork—Tools—Time—Historical Summary—On Calculations of the Numbers for Wheels and Pinions; their Proportional Sizes, Trains, etc.—Of Dial Wheels, or Motion Work—Length of Time of Going without Winding up—The Verge—The Horizontal—The Duplex—The Lever—The Chronometer—Repeating Watches—Keyless Watches—The Pendulum, or Spiral Spring—Compensation—Jewelling of Pivot Holes—Clerkenwell—Fallacies of the Trade—Incapacity of Workmen—How to Choose and Use a Watch, etc.

Algebra Self-Taught. By W. P. HIGGS, M.A., D.Sc., LL.D., Assoc. Inst. C.E., Author of 'A Handbook of the Differential Calculus,' etc. Second edition, crown 8vo, cloth, 2s. 6d.

CONTENTS:

Symbols and the Signs of Operation—The Equation and the Unknown Quantity—Positive and Negative Quantities—Multiplication—Involution—Exponents—Negative Exponents—Roots, and the Use of Exponents as Logarithms—Logarithms—Tables of Logarithms and Proportionate Parts — Transformation of System of Logarithms — Common Uses of Common Logarithms—Compound Multiplication and the Binomial Theorem—Division, Fractions, and Ratio—Continued Proportion—The Series and the Summation of the Series—Limit of Series—Square and Cube Roots—Equations—List of Formulæ, etc.

Spons' Dictionary of Engineering, Civil, Mechanical, Military, and Naval; with technical terms in French, German, Italian, and Spanish, 3100 pp., and *nearly 8000 engravings*, in super-royal 8vo, in 8 divisions, 5l. 8s. Complete in 3 vols., cloth, 5l. 5s. Bound in a superior manner, half-morocco, top edge gilt, 3 vols., 6l. 12s.

PUBLISHED BY E. & F. N. SPON. 15

Notes in Mechanical Engineering. Compiled principally for the use of the Students attending the Classes on this subject at the City of London College. By HENRY ADAMS, Mem. Inst. M.E., Mem. Inst. C.E., Mem. Soc. of Engineers. Crown 8vo, cloth, 2s. 6d.

Canoe and Boat Building: a complete Manual for Amateurs, containing plain and comprehensive directions for the construction of Canoes, Rowing and Sailing Boats, and Hunting Craft. By W. P. STEPHENS. *With numerous illustrations and 24 plates of Working Drawings.* Crown 8vo, cloth, 9s.

Proceedings of the National Conference of Electricians, Philadelphia, October 8th to 13th, 1884. 18mo, cloth, 3s.

Dynamo - Electricity, its Generation, Application, Transmission, Storage, and Measurement. By G. B. PRESCOTT. *With 545 illustrations.* 8vo, cloth, 1l. 1s.

Domestic Electricity for Amateurs. Translated from the French of E. HOSPITALIER, Editor of "L'Electricien," by C. J. WHARTON, Assoc. Soc. Tel. Eng. *Numerous illustrations.* Demy 8vo, cloth, 6s.

CONTENTS:

1. Production of the Electric Current—2. Electric Bells—3. Automatic Alarms—4. Domestic Telephones—5. Electric Clocks—6. Electric Lighters—7. Domestic Electric Lighting—8. Domestic Application of the Electric Light—9. Electric Motors—10. Electrical Locomotion—11. Electrotyping, Plating, and Gilding—12. Electric Recreations—13. Various applications—Workshop of the Electrician.

Wrinkles in Electric Lighting. By VINCENT STEPHEN. *With illustrations.* 18mo, cloth, 2s. 6d.

CONTENTS:

1. The Electric Current and its production by Chemical means—2. Production of Electric Currents by Mechanical means—3. Dynamo-Electric Machines—4. Electric Lamps—. Lead—6. Ship Lighting.

Foundations and Foundation Walls for all classes of Buildings, Pile Driving, Building Stones and Bricks, Pier and Wall construction, Mortars, Limes, Cements, Concretes, Stuccos, &c. 64 *illustrations.* By G. T. POWELL and F. BAUMAN. 8vo, cloth, 10s. 6d.

Manual for Gas Engineering Students. By D. LEE. 18mo, cloth, 1s.

Telephones, their Construction and Management.
By F. C. ALLSOP. Crown 8vo, cloth, 5s.

Hydraulic Machinery, Past and Present. A Lecture delivered to the London and Suburban Railway Officials' Association. By H. ADAMS, Mem. Inst. C.E. *Folding plate.* 8vo, sewed, 1s.

Twenty Years with the Indicator. By THOMAS PRAY, Jun., C.E., M.E., Member of the American Society of Civil Engineers. 2 vols., royal 8vo, cloth, 12s. 6d.

Annual Statistical Report of the Secretary to the Members of the Iron and Steel Association on the Home and Foreign Iron and Steel Industries in 1889. Issued June 1890. 8vo, sewed, 5s.

Bad Drains, and How to Test them; with Notes on the Ventilation of Sewers, Drains, and Sanitary Fittings, and the Origin and Transmission of Zymotic Disease. By R. HARRIS REEVES. Crown 8vo, cloth, 3s. 6d.

Well Sinking. The modern practice of Sinking and Boring Wells, with geological considerations and examples of Wells. By ERNEST SPON, Assoc. Mem. Inst. C.E., Mem. Soc. Eng., and of the Franklin Inst., etc. Second edition, revised and enlarged. Crown 8vo, cloth, 10s. 6d.

The Voltaic Accumulator: an Elementary Treatise. By ÉMILE REYNIER. Translated by J. A. BERLY, Assoc. Inst. E.E. With 62 *illustrations,* 8vo, cloth, 9s.

Ten Years' Experience in Works of Intermittent Downward Filtration. By J. BAILEY DENTON, Mem. Inst. C.E. Second edition, with additions. Royal 8vo, cloth, 5s.

Land Surveying on the Meridian and Perpendicular System. By WILLIAM PENMAN, C.E. 8vo, cloth, 8s. 6d.

The Electromagnet and Electromagnetic Mechanism. By SILVANUS P. THOMPSON, D.Sc., F.R.S. Second edition, 8vo, cloth, 15s.

Incandescent Wiring Hand-Book. By F. B. BADT,
late 1st Lieut. Royal Prussian Artillery. With 41 *illustrations and* 5 *tables.* 18mo, cloth, 4s. 6d.

A Pocket-book for Pharmacists, Medical Prac-
titioners, Students, etc., etc. (*British, Colonial, and American*). By THOMAS BAYLEY, Assoc. R. Coll. of Science, Consulting Chemist, Analyst, and Assayer, Author of a 'Pocket-book for Chemists,' 'The Assay and Analysis of Iron and Steel, Iron Ores, and Fuel,' etc., etc. Royal 32mo, boards, gilt edges, 6s.

The Fireman's Guide; a Handbook on the Care of
Boilers. By TEKNOLOG, föreningen T. I. Stockholm. Translated from the third edition, and revised by KARL P. DAHLSTROM, M.E. Second edition. Fcap. 8vo, cloth, 2s.

The Mechanician: A Treatise on the Construction
and Manipulation of Tools, for the use and instruction of Young Engineers and Scientific Amateurs, comprising the Arts of Blacksmithing and Forging; the Construction and Manufacture of Hand Tools, and the various Methods of Using and Grinding them; description of Hand and Machine Processes; Turning and Screw Cutting. By CAMERON KNIGHT, Engineer. *Containing* 1147 *illustrations*, and 397 pages of letter-press. Fourth edition, 4to, cloth, 18s.

A Treatise on Modern Steam Engines and Boilers,
including Land Locomotive, and Marine Engines and Boilers, for the use of Students. By FREDERICK COLYER, M. Inst. C.E., Mem. Inst. M.E. With 36 *plates*. 4to, cloth, 12s. 6d.

CONTENTS:

1. Introduction—2. Original Engines—3. Boilers—4. High-Pressure Beam Engines—5. Cornish Beam Engines—6. Horizontal Engines—7. Oscillating Engines—8. Vertical High-Pressure Engines—9. Special Engines—10. Portable Engines—11. Locomotive Engines—12. Marine Engines.

Steam Engine Management; a Treatise on the
Working and Management of Steam Boilers. By F. COLYER, M. Inst. C.E., Mem. Inst. M.E. New edition, 18mo, cloth, 3s. 6d.

Aid Book to Engineering Enterprise. By EWING
MATHESON, M. Inst. C.E. The Inception of Public Works, Parliamentary Procedure for Railways, Concessions for Foreign Works, and means of Providing Money, the Points which determine Success or Failure, Contract and Purchase, Commerce in Coal, Iron, and Steel, &c. Second edition, revised and enlarged, 8vo, cloth, 21s.

Pumps, Historically, Theoretically, and Practically
Considered. By P. R. BJÖRLING. With 156 *illustrations*. Crown 8vo, cloth, 7s. 6d.

The Marine Transport of Petroleum. A Book for the use of Shipowners, Shipbuilders, Underwriters, Merchants, Captains and Officers of Petroleum-carrying Vessels. By G. H. LITTLE, Editor of the 'Liverpool Journal of Commerce.' Crown 8vo, cloth, 10s. 6d.

Liquid Fuel for Mechanical and Industrial Purposes.
Compiled by E. A. BRAYLEY HODGETTS. With *wood engravings*. 8vo, cloth, 7s. 6d.

Tropical Agriculture: A Treatise on the Culture, Preparation, Commerce and Consumption of the principal Products of the Vegetable Kingdom. By P. L. SIMMONDS, F.L.S., F.R.C.I. New edition, revised and enlarged, 8vo, cloth, 21s.

Health and Comfort in House Building; or, Ventilation with Warm Air by Self-acting Suction Power. With Review of the Mode of Calculating the Draught in Hot-air Flues, and with some Actual Experiments by J. DRYSDALE, M.D., and J. W. HAYWARD, M.D. *With plates and woodcuts.* Third edition, with some New Sections, and the whole carefully Revised, 8vo, cloth, 7s. 6d.

Losses in Gold Amalgamation. With Notes on the Concentration of Gold and Silver Ores. *With six plates.* By W. MCDERMOTT and P. W. DUFFIELD. 8vo, cloth, 5s.

A Guide for the Electric Testing of Telegraph Cables.
By Col. V. HOSKIŒR, Royal Danish Engineers. Third edition, crown 8vo, cloth, 4s. 6d.

The Hydraulic Gold Miners' Manual. By T. S. G. KIRKPATRICK, M.A. Oxon. *With 6 plates.* Crown 8vo, cloth, 6s.

"We venture to think that this work will become a text-book on the important subject of which it treats. Until comparatively recently hydraulic mines were neglected. This was scarcely to be surprised at, seeing that their working in California was brought to an abrupt termination by the action of the farmers on the *débris* question, whilst their working in other parts of the world had not been attended with the anticipated success."—*The Mining World and Engineering Record.*

A Text-Book of Tanning, embracing the Preparation of all kinds of Leather. By HARRY R. PROCTOR, F.C.S., of Low Lights Tanneries. *With illustrations.* Crown 8vo, cloth, 10s. 6d.

The Arithmetic of Electricity. By T. O'CONOR
SLOANE. Crown 8vo, cloth, 4s. 6d.

The Turkish Bath: Its Design and Construction for
Public and Commercial Purposes. By R. O. ALLSOP, Architect. *With
plans and sections.* 8vo, cloth, 6s.

Earthwork Slips and Subsidences upon Public Works:
Their Causes, Prevention and Reparation. Especially written to assist
those engaged in the Construction or Maintenance of Railways, Docks,
Canals, Waterworks, River Banks, Reclamation Embankments, Drainage
Works, &c., &c. By JOHN NEWMAN, Assoc. Mem. Inst. C.E., Author
of 'Notes on Concrete,' &c. Crown 8vo, cloth, 7s. 6d.

Gas and Petroleum Engines: A Practical Treatise
on the Internal Combustion Engine. By WM. ROBINSON, M.E., Senior
Demonstrator and Lecturer on Applied Mechanics, Physics, &c., City
and Guilds of London College, Finsbury, Assoc. Mem. Inst. C.E., &c.
Numerous illustrations. 8vo, cloth, 14s.

Waterways and Water Transport in Different Countries. With a description of the Panama, Suez, Manchester, Nicaraguan,
and other Canals. By J. STEPHEN JEANS, Author of 'England's
Supremacy,' 'Railway Problems,' &c. *Numerous illustrations.* 8vo,
cloth, 14s.

A Treatise on the Richards Steam-Engine Indicator
and the Development and Application of Force in the Steam-Engine.
By CHARLES T. PORTER. Fourth Edition, revised and enlarged, 8vo,
cloth, 9s.

CONTENTS.

The Nature and Use of the Indicator:
The several lines on the Diagram.
Examination of Diagram No. 1.
Of Truth in the Diagram.
Description of the Richards Indicator.
Practical Directions for Applying and Taking Care of the Indicator.
Introductory Remarks.
Units.
Expansion.
Directions for ascertaining from the Diagram the Power exerted by the Engine.
To Measure from the Diagram the Quantity of Steam Consumed.
To Measure from the Diagram the Quantity of Heat Expended.
Of the Real Diagram, and how to Construct it.
Of the Conversion of Heat into Work in the Steam-engine.
Observations on] the several Lines of the Diagram.

Of the Loss attending the Employment of Slow-piston Speed, and the Extent to which this is Shown by the Indicator.
Of other Applications of the Indicator.
Of the use of the Tables of the Properties of Steam in Calculating the Duty of Boilers.
Introductory.
Of the Pressure on the Crank when the Connecting-rod is conceived to be of Infinite Length.
The Modification of the Acceleration and Retardation that is occasioned by the Angular Vibration of the Connecting-rod.
Method of representing the actual pressure on the crank at every point of its revolution.
The Rotative Effect of the Pressure exerted on the Crank.
The Transmitting Parts of an Engine, considered as an Equaliser of Motion.
A Ride on a Buffer-beam (Appendix).

In demy 4to, handsomely bound in cloth, *illustrated with* **220** *full page plates*,
Price 15s.

ARCHITECTURAL EXAMPLES
IN BRICK, STONE, WOOD, AND IRON.
A COMPLETE WORK ON THE DETAILS AND ARRANGEMENT OF BUILDING CONSTRUCTION AND DESIGN.

By WILLIAM FULLERTON, Architect.

Containing 220 Plates, with numerous Drawings selected from the Architecture of Former and Present Times.

The Details and Designs are Drawn to Scale, $\frac{1}{8}"$, $\frac{1}{4}"$, $\frac{1}{2}"$, *and Full size being chiefly used.*

The Plates are arranged in Two Parts. The First Part contains Details of Work in the four principal Building materials, the following being a few of the subjects in this Part:—Various forms of Doors and Windows, Wood and Iron Roofs, Half Timber Work, Porches, Towers, Spires, Belfries, Flying Buttresses, Groining, Carving, Church Fittings, Constructive and Ornamental Iron Work, Classic and Gothic Molds and Ornament, Foliation Natural and Conventional, Stained Glass, Coloured Decoration, a Section to Scale of the Great Pyramid, Grecian and Roman Work, Continental and English Gothic, Pile Foundations, Chimney Shafts according to the regulations of the London County Council, Board Schools. The Second Part consists of Drawings of Plans and Elevations of Buildings, arranged under the following heads:—Workmen's Cottages and Dwellings, Cottage Residences and Dwelling Houses, Shops, Factories, Warehouses, Schools, Churches and Chapels, Public Buildings, Hotels and Taverns, and Buildings of a general character.

All the Plates are accompanied with particulars of the Work, with Explanatory Notes and Dimensions of the various parts.

Specimen Pages, reduced from the originals.

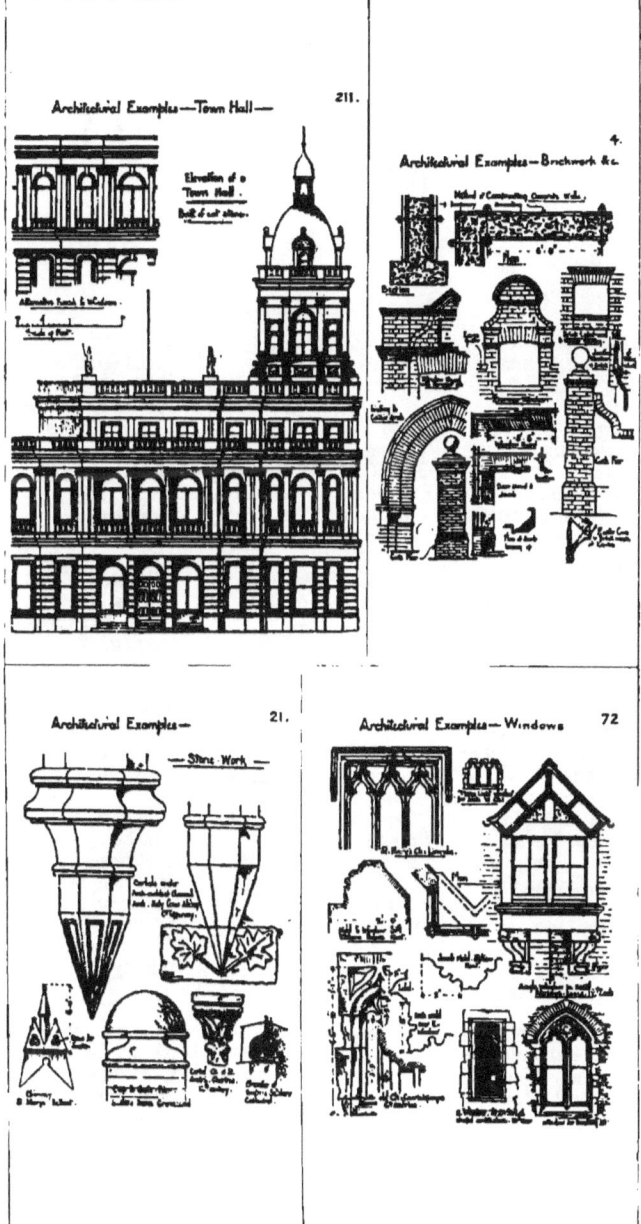

Crown 8vo, cloth, with illustrations, 5s.

WORKSHOP RECEIPTS,
FIRST SERIES.
By ERNEST SPON.

SYNOPSIS OF CONTENTS.

Bookbinding.
Bronzes and Bronzing.
Candles.
Cement.
Cleaning.
Colourwashing.
Concretes.
Dipping Acids.
Drawing Office Details.
Drying Oils.
Dynamite.
Electro-Metallurgy — (Cleaning, Dipping, Scratch-brushing, Batteries, Baths, and Deposits of every description).
Enamels.
Engraving on Wood, Copper, Gold, Silver, Steel, and Stone.
Etching and Aqua Tint.
Firework Making — (Rockets, Stars, Rains, Gerbes, Jets, Tourbillons, Candles, Fires, Lances, Lights, Wheels, Fire-balloons, and minor Fireworks).
Fluxes.
Foundry Mixtures.

Freezing.
Fulminates.
Furniture Creams, Oils, Polishes, Lacquers, and Pastes.
Gilding.
Glass Cutting, Cleaning, Frosting, Drilling, Darkening, Bending, Staining, and Painting.
Glass Making.
Glues.
Gold.
Graining.
Gums.
Gun Cotton.
Gunpowder.
Horn Working.
Indiarubber.
Japans, Japanning, and kindred processes.
Lacquers.
Lathing.
Lubricants.
Marble Working.
Matches.
Mortars.
Nitro-Glycerine.
Oils.

Paper.
Paper Hanging.
Painting in Oils, in Water Colours, as well as Fresco, House, Transparency, Sign, and Carriage Painting.
Photography.
Plastering.
Polishes.
Pottery—(Clays, Bodies, Glazes, Colours, Oils, Stains, Fluxes, Enamels, and Lustres).
Scouring.
Silvering.
Soap.
Solders.
Tanning.
Taxidermy.
Tempering Metals.
Treating Horn, Mother-o'-Pearl, and like substances.
Varnishes, Manufacture and Use of.
Veneering.
Washing.
Waterproofing.
Welding.

Besides Receipts relating to the lesser Technological matters and processes, such as the manufacture and use of Stencil Plates, Blacking, Crayons, Paste, Putty, Wax, Size, Alloys, Catgut, Tunbridge Ware, Picture Frame and Architectural Mouldings, Compos, Cameos, and others too numerous to mention.

Crown 8vo, cloth, 485 pages, with illustrations, 5s.

WORKSHOP RECEIPTS,
SECOND SERIES.

By ROBERT HALDANE.

SYNOPSIS OF CONTENTS.

Acidimetry and Alkalimetry.	Disinfectants.	Iodoform.
Albumen.	Dyeing, Staining, and Colouring.	Isinglass.
Alcohol.	Essences.	Ivory substitutes.
Alkaloids.	Extracts.	Leather.
Baking-powders.	Fireproofing.	Luminous bodies.
Bitters.	Gelatine, Glue, and Size.	Magnesia.
Bleaching.	Glycerine.	Matches.
Boiler Incrustations.	Gut.	Paper.
Cements and Lutes.	Hydrogen peroxide.	Parchment.
Cleansing.	Ink.	Perchloric acid.
Confectionery.	Iodine.	Potassium oxalate.
Copying.		Preserving.

Pigments, Paint, and Painting: embracing the preparation of *Pigments*, including alumina lakes, blacks (animal, bone, Frankfort, ivory, lamp, sight, soot), blues (antimony, Antwerp, cobalt, cæruleum, Egyptian, manganate, Paris, Péligot, Prussian, smalt, ultramarine), browns (bistre, hinau, sepia, sienna, umber, Vandyke), greens (baryta, Brighton, Brunswick, chrome, cobalt, Douglas, emerald, manganese, mitis, mountain, Prussian, sap, Scheele's, Schweinfurth, titanium, verdigris, zinc), reds (Brazilwood lake, carminated lake, carmine, Cassius purple, cobalt pink, cochineal lake, colcothar, Indian red, madder lake, red chalk, red lead, vermilion), whites (alum, baryta, Chinese, lead sulphate, white lead—by American, Dutch, French, German, Kremnitz, and Pattinson processes, precautions in making, and composition of commercial samples—whiting, Wilkinson's white, zinc white), yellows (chrome, gamboge, Naples, orpiment, realgar, yellow lakes); *Paint* (vehicles, testing oils, driers, grinding, storing, applying, priming, drying, filling, coats, brushes, surface, water-colours, removing smell, discoloration; miscellaneous paints—cement paint for carton-pierre, copper paint, gold paint, iron paint, lime paints, silicated paints, steatite paint, transparent paints, tungsten paints, window paint, zinc paints); *Painting* (general instructions, proportions of ingredients, measuring paint work; carriage painting—priming paint, best putty, finishing colour, cause of cracking, mixing the paints, oils, driers, and colours, varnishing, importance of washing vehicles, re-varnishing, how to dry paint; woodwork painting).

Crown 8vo, cloth, 480 pages, with 183 illustrations, 5s.

WORKSHOP RECEIPTS,

THIRD SERIES.

By C. G. WARNFORD LOCK.

Uniform with the First and Second Series.

SYNOPSIS OF CONTENTS.

Alloys.	Indium.	Rubidium.
Aluminium.	Iridium.	Ruthenium.
Antimony.	Iron and Steel.	Selenium.
Barium.	Lacquers and Lacquering.	Silver.
Beryllium.	Lanthanum.	Slag.
Bismuth.	Lead.	Sodium.
Cadmium.	Lithium.	Strontium.
Cæsium.	Lubricants.	Tantalum.
Calcium.	Magnesium.	Terbium.
Cerium.	Manganese.	Thallium.
Chromium.	Mercury.	Thorium.
Cobalt.	Mica.	Tin.
Copper.	Molybdenum.	Titanium.
Didymium.	Nickel.	Tungsten.
Electrics.	Niobium.	Uranium.
Enamels and Glazes.	Osmium.	Vanadium.
Erbium.	Palladium.	Yttrium.
Gallium.	Platinum.	Zinc.
Glass.	Potassium.	Zirconium.
Gold.	Rhodium.	

WORKSHOP RECEIPTS,
FOURTH SERIES,
DEVOTED MAINLY TO HANDICRAFTS & MECHANICAL SUBJECTS.

By C. G. WARNFORD LOCK.

250 Illustrations, with Complete Index, and a General Index to the Four Series, 5s.

Waterproofing — rubber goods, cuprammonium processes, miscellaneous preparations.
Packing and Storing articles of delicate odour or colour, of a deliquescent character, liable to ignition, apt to suffer from insects or damp, or easily broken.
Embalming and Preserving anatomical specimens.
Leather Polishes.
Cooling Air and Water, producing low temperatures, making ice, cooling syrups and solutions, and separating salts from liquors by refrigeration.
Pumps and Siphons, embracing every useful contrivance for raising and supplying water on a moderate scale, and moving corrosive, tenacious, and other liquids.
Desiccating—air- and water-ovens, and other appliances for drying natural and artificial products.
Distilling—water, tinctures, extracts, pharmaceutical preparations, essences, perfumes, and alcoholic liquids.
Emulsifying as required by pharmacists and photographers.
Evaporating—saline and other solutions, and liquids demanding special precautions.
Filtering—water, and solutions of various kinds.
Percolating and Macerating.
Electrotyping.
Stereotyping by both plaster and paper processes.
Bookbinding in all its details.
Straw Plaiting and the fabrication of baskets, matting, etc.
Musical Instruments—the preservation, tuning, and repair of pianos, harmoniums, musical boxes, etc.
Clock and Watch Mending—adapted for intelligent amateurs.
Photography—recent development in rapid processes, handy apparatus, numerous recipes for sensitizing and developing solutions, and applications to modern illustrative purposes.

Crown 8vo, cloth, with 373 illustrations, price 5s.

WORKSHOP RECEIPTS,
FIFTH SERIES.
By C. G. WARNFORD LOCK, F.L.S.

Containing many new Articles, as well as additions to Articles included in the previous Series, as follows, viz. :—

Anemometers.
Barometers, How to make.
Boat Building.
Camera Lucida, How to use.
Cements and Lutes.
Cooling.
Copying.
Corrosion and Protection of Metal Surfaces.
Dendrometer, How to use.
Desiccating.
Diamond Cutting and Polishing. Electrics. New Chemical Batteries, Bells, Commutators, Galvanometers, Cost of Electric Lighting, Microphones, Simple Motors, Phonogram and Graphophone, Registering Apparatus, Regulators, Electric Welding and Apparatus, Transformers.
Evaporating.
Explosives.
Filtering.
Fireproofing, Buildings, Textile Fabrics.
Fire-extinguishing Compounds and Apparatus.
Glass Manipulating. Drilling, Cutting, Breaking, Etching, Frosting, Powdering, &c.

Glass Manipulations for Laboratory Apparatus.
Labels. Lacquers.
Illuminating Agents.
Inks. Writing, Copying, Invisible, Marking, Stamping.
Magic Lanterns, their management and preparation of slides.
Metal Work. Casting Ornamental Metal Work, Copper Welding, Enamels for Iron and other Metals, Gold Beating, Smiths' Work.
Modelling and Plaster Casting.
Netting.
Packing and Storing. Acids, &c.
Percolation.
Preserving Books.
Preserving Food, Plants, &c.
Pumps and Syphons for various liquids.
Repairing Books.
Rope Tackle.
Stereotyping.
Taps, Various.
Tobacco Pipe Manufacture.
Tying and Splicing Ropes.
Velocipedes, Repairing.
Walking Sticks.
Waterproofing.

NOW COMPLETE.

With nearly 1500 *illustrations*, in super-royal 8vo, in 5 Divisions, cloth. Divisions 1 to 4, 13*s.* 6*d.* each ; Division 5, 17*s.* 6*d.* ; or 2 vols., cloth, £3 10*s.*

SPONS' ENCYCLOPÆDIA
OF THE
INDUSTRIAL ARTS, MANUFACTURES, AND COMMERCIAL PRODUCTS.

EDITED BY C. G. WARNFORD LOCK, F.L.S.

Among the more important of the subjects treated of, are the following :—

Acids, 207 pp. 220 figs.
Alcohol, 23 pp. 16 figs.
Alcoholic Liquors, 13 pp.
Alkalies, 89 pp. 78 figs.
Alloys.   Alum.
Asphalt.   Assaying.
Beverages, 89 pp. 29 figs.
Blacks.
Bleaching Powder, 15 pp.
Bleaching, 51 pp. 48 figs.
Candles, 18 pp. 9 figs.
Carbon Bisulphide.
Celluloid, 9 pp.
Cements.   Clay.
Coal-tar Products, 44 pp. 14 figs.
Cocoa, 8 pp.
Coffee, 32 pp. 13 figs.
Cork, 8 pp. 17 figs.
Cotton Manufactures, 62 pp. 57 figs.
Drugs, 38 pp.
Dyeing and Calico Printing, 28 pp. 9 figs.
Dyestuffs, 16 pp.
Electro-Metallurgy, 13 pp.
Explosives, 22 pp. 33 figs.
Feathers.
Fibrous Substances, 92 pp. 79 figs.
Floor-cloth, 16 pp. 21 figs.
Food Preservation, 8 pp.
Fruit, 8 pp.

Fur, 5 pp.
Gas, Coal, 8 pp.
Gems.
Glass, 45 pp. 77 figs.
Graphite, 7 pp.
Hair, 7 pp.
Hair Manufactures.
Hats, 26 pp. 26 figs.
Honey.   Hops.
Horn.
Ice, 10 pp. 14 figs.
Indiarubber Manufactures, 23 pp. 17 figs.
Ink, 17 pp.
Ivory.
Jute Manufactures, 11 pp., 11 figs.
Knitted Fabrics — Hosiery, 15 pp. 13 figs.
Lace, 13 pp. 9 figs.
Leather, 28 pp. 31 figs.
Linen Manufactures, 16 pp. 6 figs.
Manures, 21 pp. 30 figs.
Matches, 17 pp. 38 figs.
Mordants, 13 pp.
Narcotics, 47 pp.
Nuts, 10 pp.
Oils and Fatty Substances, 125 pp.
Paint.
Paper, 26 pp. 23 figs.
Paraffin, 8 pp. 6 figs.
Pearl and Coral, 8 pp.
Perfumes, 10 pp.

Photography, 13 pp. 20 figs.
Pigments, 9 pp. 6 figs.
Pottery, 46 pp. 57 figs.
Printing and Engraving, 20 pp. 8 figs.
Rags.
Resinous and Gummy Substances, 75 pp. 16 figs.
Rope, 16 pp. 17 figs.
Salt, 31 pp. 23 figs.
Silk, 8 pp.
Silk Manufactures, 9 pp. 11 figs.
Skins, 5 pp.
Small Wares, 4 pp.
Soap and Glycerine, 39 pp. 45 figs.
Spices, 16 pp.
Sponge, 5 pp.
Starch, 9 pp. 10 figs.
Sugar, 155 pp. 134 figs.
Sulphur.
Tannin, 18 pp.
Tea, 12 pp.
Timber, 13 pp.
Varnish, 15 pp.
Vinegar, 5 pp.
Wax, 5 pp.
Wool, 2 pp.
Woollen Manufactures, 58 pp. 39 figs.

MECHANICAL MANIPULATION.

THE MECHANICIAN:
A TREATISE ON THE CONSTRUCTION AND MANIPULATION OF TOOLS, FOR THE USE AND INSTRUCTION OF YOUNG ENGINEERS AND SCIENTIFIC AMATEURS;

Comprising the Arts of Blacksmithing and Forging the Construction and Manufacture of Hand Tools, and the various Methods of Using and Grinding them; the Construction of Machine Tools, and how to work them; Turning and Screw-cutting; the various details of setting out work, &c., &c.

By CAMERON KNIGHT, Engineer.

96 4to plates, containing 1147 illustrations, and 397 pages of letterpress, second edition, reprinted from the first, 4to, cloth, 18s.

Of the six chapters constituting the work, the first is devoted to forging; in which the fundamental principles to be observed in making forged articles of every class are stated, giving the proper relative positions for the constituent fibres of each article, the mode of selecting proper quantities of material, steam-hammer operations, shaping-moulds, and the manipulations resorted to for shaping the component masses to the intended forms.

Engineers' tools and their construction are next treated, because they must be used during all operations described in the remaining chapters, the author thinking that the student should first acquire knowledge of the apparatus which he is supposed to be using in the course of the processes given in Chapters 4, 5, and 6. In the fourth chapter planing and lining are treated, because these are the elements of machine-making in general. The processes described in this chapter are those on which all accuracy of fitting and finishing depend. The next chapter, which treats of shaping and slotting, the author endeavours to render comprehensive by giving the hand-shaping processes in addition to the machine-shaping.

In many cases hand-shaping is indispensable, such as sudden breakage, operations abroad, and on board ship, also for constructors having a limited number of machines. Turning and screw-cutting occupy the last chapter. In this, the operations for lining, centering, turning, and screw-forming are detailed and their principles elucidated.

The Mechanician is the result of the author's experience in engine making during twenty years; and he has concluded that, however retentive the memory of a machinist might be, it would be convenient for him to have a book of primary principles and processes to which he could refer with confidence.

PUBLISHED BY E. & F. N. SPON.

JUST PUBLISHED.

In demy 8vo, cloth, 600 pages, and 1420 Illustrations, 6s.

SPONS'
MECHANICS' OWN BOOK;
A MANUAL FOR HANDICRAFTSMEN AND AMATEURS.

CONTENTS.

Mechanical Drawing—Casting and Founding in Iron, Brass, Bronze, and other Alloys—Forging and Finishing Iron—Sheetmetal Working—Soldering, Brazing, and Burning—Carpentry and Joinery, embracing descriptions of some 400 Woods, over 200 Illustrations of Tools and their uses, Explanations (with Diagrams) of 116 joints and hinges, and Details of Construction of Workshop appliances, rough furniture, Garden and Yard Erections, and House Building—Cabinet-Making and Veneering — Carving and Fretcutting — Upholstery — Painting, Graining, and Marbling — Staining Furniture, Woods, Floors, and Fittings—Gilding, dead and bright, on various grounds—Polishing Marble, Metals, and Wood—Varnishing—Mechanical movements, illustrating contrivances for transmitting motion—Turning in Wood and Metals—Masonry, embracing Stonework, Brickwork, Terracotta and Concrete—Roofing with Thatch, Tiles, Slates, Felt, Zinc, &c.—Glazing with and without putty, and lead glazing—Plastering and Whitewashing— Paper-hanging— Gas-fitting—Bell-hanging, ordinary and electric Systems — Lighting — Warming — Ventilating — Roads, Pavements, and Bridges — Hedges, Ditches, and Drains — Water Supply and Sanitation—Hints on House Construction suited to new countries.

E. & F. N. SPON, 125, Strand, London.
New York: 12, Cortlandt Street.

SPONS' DICTIONARY OF ENGINEERING,

CIVIL, MECHANICAL, MILITARY, & NAVAL,

WITH

Technical Terms in French, German, Italian, and Spanish.

In 97 numbers, Super-royal 8vo, containing 3132 *printed pages* and 7414 *engravings*. Any number can be had separate: Nos. 1 to 95 1*s*. each, post free; Nos. 96, 97, 2*s*., post free. *See also page* 112.

COMPLETE LIST OF ALL THE SUBJECTS:

	Nos.		Nos.
Abacus	1	Barrage	8 and 9
Adhesion	1	Battery	9 and 10
Agricultural Engines	1 and 2	Bell and Bell-hanging	10
Air-Chamber	2	Belts and Belting	10 and 11
Air-Pump	2	Bismuth	11
Algebraic Signs	2	Blast Furnace	11 and 12
Alloy	2	Blowing Machine	12
Aluminium	2	Body Plan	12 and 13
Amalgamating Machine	2	Boilers	13, 14, 15
Ambulance	2	Bond	15 and 16
Anchors	2	Bone Mill	16
Anemometer	2 and 3	Boot-making Machinery	16
Angular Motion	3 and 4	Boring and Blasting	16 to 19
Angle-iron	3	Brake	19 and 20
Angle of Friction	3	Bread Machine	20
Animal Charcoal Machine	4	Brewing Apparatus	20 and 21
Antimony, 4; Anvil	4	Brick-making Machines	21
Aqueduct, 4; Arch	4	Bridges	21 to 28
Archimedean Screw	4	Buffer	28
Arming Press	4 and 5	Cables	28 and 29
Armour, 5; Arsenic	5	Cam, 29; Canal	29
Artesian Well	5	Candles	29 and 30
Artillery, 5 and 6; Assaying	6	Cement, 30; Chimney	30
Atomic Weights	6 and 7	Coal Cutting and Washing Machinery	31
Auger, 7; Axles	7	Coast Defence	31, 32
Balance, 7; Ballast	7	Compasses	32
Bank Note Machinery	7	Construction	32 and 33
Barn Machinery	7 and 8	Cooler, 34; Copper	34
Barker's Mill	8	Cork-cutting Machine	34
Barometer, 8; Barracks	8		

PUBLISHED BY E. & F. N. SPON.

	Nos.		Nos.
Corrosion	34 and 35	Isomorphism, 68 ; Joints	.. 68
Cotton Machinery	.. 35	Keels and Coal Shipping	68 and 69
Damming	35 to 37	Kiln, 69 ; Knitting Machine	.. 69
Details of Engines	37, 38	Kyanising	.. 69
Displacement	.. 38	Lamp, Safety	69, 70
Distilling Apparatus	38 and 39	Lead	.. 70
Diving and Diving Bells	.. 39	Lifts, Hoists	70, 71
Docks	39 and 40	Lights, Buoys, Beacons	71 and 72
Drainage	40 and 41	Limes, Mortars, and Cements	.. 72
Drawbridge	.. 41	Locks and Lock Gates	72, 73
Dredging Machine	.. 41	Locomotive	.. 73
Dynamometer	41 to 43	Machine Tools	73, 74
Electro-Metallurgy	43, 44	Manganese	.. 74
Engines, Varieties	44, 45	Marine Engine	74 and 75
Engines, Agricultural	1 and 2	Materials of Construction	75 and 76
Engines, Marine	74, 75	Measuring and Folding	.. 76
Engines, Screw	89, 90	Mechanical Movements	76, 77
Engines, Stationary	91, 92	Mercury, 77 ; Metallurgy	.. 77
Escapement	45, 46	Meter	77, 78
Fan	.. 46	Metric System	.. 78
File-cutting Machine	.. 46	Mills	78, 79
File-arms	46, 47	Molecule, 79 ; Oblique Arch	.. 79
Flax Machinery	47, 48	Ores, 79, 80 ; Ovens	.. 80
Float Water-wheels	.. 48	Over-shot Water-wheel	80, 81
Forging	.. 48	Paper Machinery	.. 81
Founding and Casting	48 to 50	Permanent Way	81, 82
Friction, 50 ; Friction, Angle of	3	Piles and Pile-driving	82 and 83
Fuel, 50 ; Furnace	50, 51	Pipes	83, 84
Fuze, 51 ; Gas	.. 51	Planimeter	.. 84
Gearing	51, 52	Pumps	84 and 85
Gearing Belt	10, 11	Quarrying	.. 85
Geodesy	52 and 53	Railway Engineering	85 and 86
Glass Machinery	.. 53	Retaining Walls	.. 86
Gold, 53, 54 ; Governor	.. 54	Rivers, 86, 87 ; Rivetted Joint	.. 87
Gravity, 54 ; Grindstone	.. 54	Roads	87, 88
Gun-carriage, 54 ; Gun Metal	.. 54	Roofs	88, 89
Gunnery	54 to 56	Rope-making Machinery	.. 89
Gunpowder	.. 56	Scaffolding	.. 89
Gun Machinery	56, 57	Screw Engines	89, 90
Hand Tools	57, 58	Signals, 90 ; Silver	90, 91
Hanger, 58 ; Harbour	.. 58	Stationary Engine	91, 92
Haulage, 58, 59 ; Hinging	.. 59	Stave-making & Cask Machinery	92
Hydraulics and Hydraulic Machinery	59 to 63	Steel, 92 ; Sugar Mill	92, 93
		Surveying and Surveying Instruments	93, 94
Ice-making Machine	.. 63		
India-rubber	.. 63	Telegraphy	94, 95
Indicator	63 and 64	Testing, 95 ; Turbine	.. 95
Injector	.. 64	Ventilation	95, 96, 97
Iron	64 to 67	Waterworks	96, 97
Iron Ship Building	.. 67	Wood-working Machinery	96, 97
Irrigation	67 and 68	Zinc	97

In super-royal 8vo, 1168 pp., *with* 2400 *illustrations*, in 3 Divisions, cloth, price 13s. 6d. each ; or 1 vol., cloth, 2l. ; or half-morocco, 2l. 8s.

A SUPPLEMENT

TO

SPONS' DICTIONARY OF ENGINEERING.

EDITED BY ERNEST SPON, MEMB. SOC. ENGINEERS.

Abacus, Counters, Speed Indicators, and Slide Rule.
Agricultural Implements and Machinery.
Air Compressors.
Animal Charcoal Machinery.
Antimony.
Axles and Axle-boxes.
Barn Machinery.
Belts and Belting.
Blasting. Boilers.
Brakes.
Brick Machinery.
Bridges.
Cages for Mines.
Calculus, Differential and Integral.
Canals.
Carpentry.
Cast Iron.
Cement, Concrete, Limes, and Mortar.
Chimney Shafts.
Coal Cleansing and Washing.
Coal Mining.
Coal Cutting Machines.
Coke Ovens. Copper.
Docks. Drainage.
Dredging Machinery.
Dynamo - Electric and Magneto-Electric Machines.
Dynamometers.
Electrical Engineering, Telegraphy, Electric Lighting and its practical details, Telephones
Engines, Varieties of.
Explosives. Fans.
Founding, Moulding and the practical work of the Foundry.
Gas, Manufacture of.
Hammers, Steam and other Power.
Heat. Horse Power.
Hydraulics.
Hydro-geology.
Indicators. Iron.
Lifts, Hoists, and Elevators.
Lighthouses, Buoys, and Beacons.
Machine Tools.
Materials of Construction.
Meters.
Ores, Machinery and Processes employed to Dress.
Piers.
Pile Driving.
Pneumatic Transmission.
Pumps.
Pyrometers.
Road Locomotives.
Rock Drills.
Rolling Stock.
Sanitary Engineering.
Shafting.
Steel.
Steam Navvy.
Stone Machinery.
Tramways.
Well Sinking.

www.ingramcontent.com/pod-product-compliance
Lightning Source LLC
Chambersburg PA
CBHW020251170426
43202CB00008B/319